DESIGN OF MACHINE AND STRUCTURAL PARTS

DESIGN OF MACHINE AND STRUCTURAL PARTS

Kurt M. Marshek
Department of Mechanical Engineering
University of Texas at Austin
Austin, Texas

A WILEY-INTERSCIENCE PUBLICATION

JOHN WILEY & SONS

NEW YORK   CHICHESTER   BRISBANE   TORONTO   SINGAPORE

Copyright © 1987 by John Wiley & Sons, Inc.

All rights reserved. Published simultaneously in Canada.

Reproduction or translation of any part of this work
beyond that permitted by Section 107 or 108 of the
1976 United States Copyright Act without the permission
of the copyright owner is unlawful. Requests for
permission or further information should be addressed to
the Permissions Department, John Wiley & Sons, Inc.

Library of Congress Cataloging in Publication Data:
Marshek, Kurt M.
    Design of machine and structural parts.

    "A Wiley-Interscience publication."
    Bibliography: p.
    Includes index.
    1. Machinery—Design. 2. Structural design.
I. Title.
TJ230.M32 1987   621.8′15   87-6286
ISBN 0-471-84996-0

Printed in the United States of America

10   9   8   7   6   5   4   3   2   1

To Nicki and Kelly

PREFACE

This book was written for students of product, machine, and structural design. Since design involves geometry, the ability to sketch and draw the shapes and forms of parts is requisite. This book concentrates on presenting the principles of form design and how specific forms and loading systems affect part stresses and rigidities.

This book is intended to supplement the study of component design analysis. It is assumed that the user has had basic courses in statics and strength of materials. The objective is to provide numerous examples and illustrations for improving the strength and minimizing the weight of parts. Once learned, the principles presented here can be used for the efficient form synthesis of machine and structural parts. The book can also serve as a review of strength of materials and will allow the student to become a specialist in the initial synthesis phase of design.

The chapters depend upon, and largely follow one another, building on previously developed principles and ideas. The first three chapters of the book serve to review as well as extend the basic background. The next few chapters deal with the application of some fundamental principles to specific parts and structures. The final chapter deals with the effect of shape and loading system on rigidity. For a person with a background in strength of materials, the chapters could be studied in almost any order. The principles presented are fundamental to the form design of machine and structural parts and are described in detail.

Although much of the first three chapters is a review of statics and strength of materials, several sections are of particular importance. In Chapter 1, Section 1-3 deals with free body diagrams. Most engineers recognize the importance of being able to draw free-body diagrams and determine loads. Clearly, if the loading on a part cannot be properly determined, the consequent design analysis is meaningless. Section 1-4 presents the idea of being able to visualize the flow of force through a machine part, a technique that is invaluable for identifying the locatı̇ ıs of maximum stresses. Section 1-7 emphasizes the relative strengths of stress patterns, and Section 1-10 reviews St. Venant's principle and demonstrates how rapidly even a knife edge load distribution will spread and become an almost uniform load distribution.

Chapter 2 studies in detail the efficient stress patterns of tension, compression, and uniform shear and discusses the inefficient stress patterns of bending, transverse shear stress, and spot contact.

Chapter 3 reviews the concept of a spring constant and then demonstrates that the efficient stress patterns are related to rigidity and that the inefficient stress patterns are related to flexibility. Section 3-5 gives design recommendations for increasing the rigidity of machine and structural parts. Designing for uniform stress produces rigid parts.

Chapter 4 presents the general principle for form design, which is "design for uniform stress." The principles that follow from the general principle are presented in Sections 4-2 to 4-6: tetrahedron-triangle; uniform shear or hollow shaft; forcing load; mating surface; and load-lever principle. Once understood, these principles are easy to apply in the design synthesis process.

Chapter 5 discusses tension, compression, and bending. Section 5-4 reveals configurations that are highly efficient strength-to-weight ratio shapes for bending. Section 5-5 points out the merits of the sandwich construction for use in resisting bending moments.

Chapter 6 shows how the membrane analogy can be used in designing parts loaded in torsion. Studied are the relative strength of circular vs. noncircular cross-sectional shapes, inward vs. outward protruding corners, and closed vs. open cross sections.

Chapter 7 presents a study of contact stresses. Section 7-1 begins with a discussion of Hertz equations for the general case of contact between two elastic bodies. Section 7-2 looks at contact stresses in parallel cylinders. Section 7-3 discusses material methods useful for determining contact stresses for both Hertzian and conformal contact. Section 7-4 studies contact stresses in indented strips and slabs, and Section 7-5 presents work done by Peter Engel at IBM on contact stresses in type characters. The remaining sections of Chapter 7 look at gear tooth form optimization to reduce wear and contact stresses as well as press fit shaft parts.

Chapter 8 presents buckling conceptually as a shift from an efficient to an inefficient stress pattern. For example, in column buckling there is a shift from compression to bending. Section 8-2 looks at buckling in components. Section 8-3 gives critical lengths of members for buckling by compression and torsion. The remaining sections present methods for preventing buckling and discuss the design of lightweight frames.

Chapter 9 presents the analysis for axial and torsional impact using the energy method. The energy absorbing ability is greatest for parts stressed uniformly throughout. Efficient stress patterns are preferred to inefficient patterns. Ideas are presented for improving the impact design of components. Design considerations are given for minimizing the effects of impact loads.

Chapters 10 and 11 deal with the design of joint elements. Joints can be thought of as breaks in the continuity of machines or structures. Joints are needed to efficiently manufacture and assemble machines. Principles are presented for part design and for increasing the strength-to-weight ratio efficiency of joints.

Additional principles are presented for increasing body and joint efficiency in Chapter 12. Discussed are the use of supplementary structural shapes, floating or semifloating parts, and the elastic matching and the shape refinement principles.

The final chapter, Chapter 13, presents the principle of relative stiffness. Several examples to illustrate the principle are presented: (1) load distribution in modular belting; (2) load in a three-column concrete structure; and (3) load distribution in chains and sprockets, timing belts, and threaded connectors. Also presented is a discussion of the load shift phenomenon.

The principles presented in this book are important, but they will not be sufficient to solve a complete engineering problem. It is expected that a thorough understanding of the ideas presented will guide the reader in producing well-formed machine and structural parts, components, and structures. The engineering know-how, ingenuity, insight, and imagination acquired through many years of engineering experience can be supplemented with the principles presented. Although computer power has increased and costs have decreased drastically, the understanding of the principles of form design will reduce the time and expense in producing efficient designs.

KURT M. MARSHEK

Austin, Texas
August 1987

ACKNOWLEDGMENTS

This book evolved at the beginning from classroom outlines developed by Professor Walter L. Starkey of The Ohio State University, under whom I studied over 15 years ago. My interest in the area of mechanical engineering design was influenced by the fact that I studied this field under Professor Starkey, an outstanding engineer, teacher, and gentleman whom I very much admire. Professor Starkey's classes were an introduction to a totally different way of thinking on problems ranging from design to assessing the value of free speech. (I would also be amiss if I did not mention the influence of studying Strength of Materials under Professor Ralph I. Stephens while he was teaching at the University of Wisconsin.)

After graduate study at Ohio State, I introduced the principles of form design into the design classes at The University of Connecticut. Unknown to me at the time, Professor Terry E. Shoup, a classmate of mine at Ohio State, was doing the same at Rutgers University. In 1976, we both had joined the faculty at the University of Houston, and there decided to develop and teach a short course covering the principles of form design.

Not being able to interest Professor Starkey nor Professor Shoup in writing a form design book, in 1981 I decided to undertake the task myself. Fortunately, I have had help from numerous students: David Tso, Shiran Nanayakkara, Hsien-Heng Chen, Raul Longoria, Monica Gonzalez, and Srikanth Kannapan. My goal was to simply present and illustrate, in the most efficient manner, the concepts of form design.

To rephrase the words of Professor Robert Juvinall:* "While every effort has been made to insure the accuracy and the conformity with good engineering practice of the material in this book, there is no guarantee, stated or implied, that machines or structural parts designed based on the information provided in this text will be in all cases proper and safe. Design is sufficiently complex that its actual practice should always take advantage of the literature in a specific area involved, the background of experience, and most important, appropriate experiments to verify proper and safe performance."

*Juvinall, R. C., **Fundamentals of Machine Component Design,** John Wiley & Sons, New York, 1983.

CONTENTS

1

INTRODUCTION TO FORM DESIGN

Machines and structures are monolithic assemblages comprised of body and joint elements. Generally, when a body or joint is subjected to loads, efficient and inefficient stress patterns are introduced. The problem is to be able to recognize what forms and shapes cause which stress patterns and to apply this information to produce an improved design.

1-1 GIVEN INFORMATION

The designer usually has a definition of the problem that must be solved as well as knowledge of the design considerations listed in Table 1-1 before beginning to synthesize the size and shape of a machine or structural part. The designer generally needs to know: (1) the *performance* requirements—the magnitude and location of the applied forces and the motions of the part; (2) the required *life;* (3) the *cost* involved in producing or purchasing the part; and (4) the *constraints*—space, weight, fabrication, materials, aesthetics, compatibility with other parts, and compatibility with the environment.

The information needed, for example, to select chain and sprocket size and shape for a drive system is the (1) power to be transmitted, (2) angular speed of the driven shaft, (3) desired speed ratio, (4) source and type of power, (5) available space for the drive, (6) shaft diameters, (7) type of equipment to be driven, (8) operating environment, and (9) desired life of the installation.

TABLE 1-1. Design Considerations

1. Capacity (power, load, thermal)
2. Motion (kinematics, vibration, dynamics, controllability)
3. Interfaces (appearance, space limits, load type(s), environmental compatibility)
4. Cost (initial, operating)
5. Life
6. Reliability
7. Safety
8. Noise
9. Availability of components
10. Producibility
11. Maintainability
12. Geometry (size, shape)
13. Rigidity
14. Elastic stability (buckling)
15. Weight
16. Materials (strength, cost, availability, modulus of elasticity, toughness)
17. Uncertainties (load, environment, cost, material)

1-2 MACHINE ASSEMBLAGE

A machine consists of an assemblage of monolithic parts, each of which, in general, has a *body* and several *joint* elements. Figure 1-1 identifies joint and body elements in a chain-drive system. Figure 1-2 shows joint and body elements for a channel beam whose ends are welded onto immovable supports.

1-3 FREE-BODY DIAGRAMS

A body that is composed of one or several components can be analyzed by separating it into parts. The behavior of each portion of the body can then be isolated and studied. When a portion is removed from the rest of the body, the internal

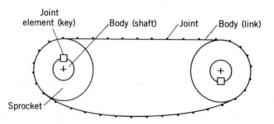

FIGURE 1-1. Chain-sprocket drive system showing body and joint elements.

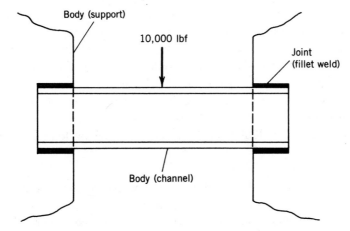

FIGURE 1-2. Channel beam welded at ends and loaded in the middle.

forces and moments that exist at the boundary between the removed and remaining portions of the body are redefined as external forces and moments (see Fig. 1-3). Diagrams of bodies or body portions are called free-body diagrams. More specifically, a *free-body diagram* is a drawing or a sketch of a body (or part of a body) that shows all the forces from the surroundings acting on that body. The forces could be caused by gravitational attraction, centrifugal acceleration, magnetic repulsion or attraction, or another body.

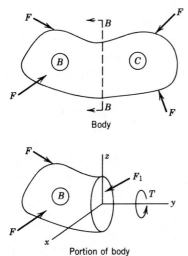

FIGURE 1-3. Construction of a free-body diagram

Even the most intricate systems can be analyzed by constructing and studying free-body diagrams of the system's various parts. By reducing the system into component subsystems, we can determine the state of each isolated subsystem with free-body diagrams. The behavior of the system as a whole can then be described by the combined effects of its parts. Free-body diagrams are helpful in facilitating solutions to complicated problems and in determining quantitative relationships between parameters.

As stated by Shigley [26]: Using free-body diagrams for force analysis serves the following purposes:

1. The diagram establishes the directions of reference axes; provides a place to record the dimensions of the subsystem and the magnitudes and directions of the known forces, and helps in assuming the directions of unknown forces.

2. The diagram simplifies an individual's thinking because it provides a place to store one thought while proceeding to the next.

3. The diagram provides a means of communicating a person's thoughts clearly and unambiguously to other people.

4. The diagram aids in understanding all facets of the problem. Careful and complete construction of the diagram clarifies fuzzy thinking by bringing out various points that are not always apparent in the statement or in the geometry of the total problem.

5. The diagram helps in the planning of a logical attack on the problem and in setting up the mathematical relations.

6. The diagram helps in recording progress in the solution and in illustrating the methods used.

As an example of free-body diagrams, consider Figure 1-4 which shows two smooth spheres in contact with each other and placed inside a cylinder with smooth walls. A free-body diagram can be sketched of each part by showing the shape of the part and all the external forces acting on each part. Diagrams are shown in Figure 1-5.

As Shames [48] points out, we might attempt to consider a portion of the container as a free body as shown in Figure 1-5, but even if this diagram did clearly depict a body (which it does not), it would not qualify as a free body since all the forces acting on the body have not been shown. The force acting on the free-body diagram of the container depends on how the container is supported. If the container is resting on a flat surface, then the supporting reaction will be distributed over the total contact area. If it is resting on two wedge supports, then there will be two concentrated forces acting on the bottom surface of the container.

Figure 1-6 shows an assembly with a hinged end at *A* and a simply supported end at *C*. The frame suspends a 1000 lbf load at *B*. Figure 1-6*c* is a free-body

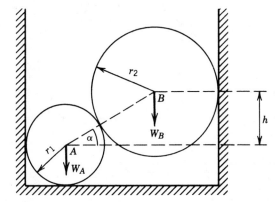

FIGURE 1-4. Two spheres in a cylindrical container with unknown supports. [From Irving H. Shames, *Engineering Mechanics: Statics and Dynamics*, © 1960, pp. 81, 82, 84, 85, 86. Reprinted by permission of Prentice-Hall, Inc., Englewood Cliffs, N.J.]

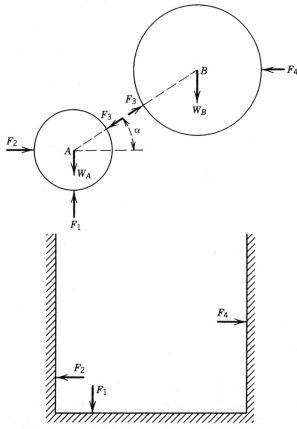

FIGURE 1-5. Free-body diagrams of spheres and incorrect free-body diagram of container. [From Irving H. Shames, *Engineering Mechanics: Statics and Dynamics*, © 1960, pp. 81, 82, 84, 85, 86. Reprinted by permission of Prentice-Hall, Inc., Englewood Cliffs, N.J.]

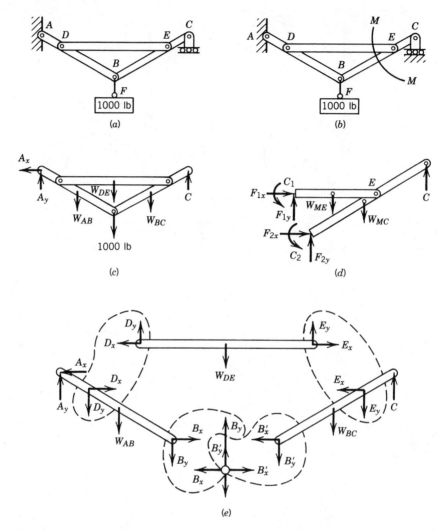

FIGURE 1-6. Free-body diagram for a pinned assembly. [From Irving H. Shames, *Engineering Mechanics: Statics and Dynamics*, © 1960, pp. 81, 82, 84, 85, 86. Reprinted by permission of Prentice-Hall, Inc., Englewood Cliffs, N.J.]

diagram of the external forces acting on the frame. Figure 1-6d is a free-body diagram involving an interior section along arc MM, shown in Figure 1-6b. Note that the internal forces and moments within the frame at the interface of the section are redefined as external moments and forces in the subsystem depicted in Figure 1-6d. Each component of the assembly can be separated and isolated to determine the loading effect on the various parts. Figure 1-6e depicts free-body diagrams of the isolated members of the assembly. The internal forces of the assembly are

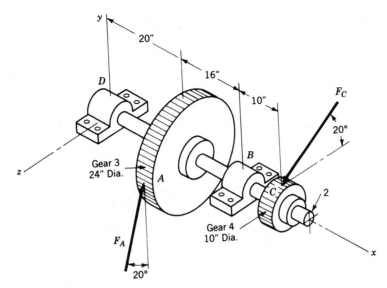

FIGURE 1-7. Shaft with two spur gears. [From Ref. 49, p. 709.]

represented as the external forces of the system's components. Note that equal and opposite reactions exist at the common joints of the individual members.

As a final example, we draw a free-body diagram for a shaft and the two gears mounted on it. The resultant gear force $F_A = 600$ lbf acts at an angle of 20° from the y axis of the overhanging countershaft shown in Figure 1-7. The bearing forces at B and D are shown as the reacting forces in Figure 1-8. Figure 1-8 is a correct free-body diagram if the weight of the gears and shaft can be neglected.

Equations of Force Equilibrium

If a body or a part of a body (as shown by a free-body diagram) is not accelerating, the summation of forces in the directions of any three orthogonal axes and the summation of moments about these axes must be zero and the body is said to be in *static equilibrium*. If the body is accelerating, then inertia forces and/or torques can be associated with the accelerations. If the inertia forces and torques are added to the original loads, then the body can be considered to be in equilibrium (d'Alembert's principle) and the summation of forces and moments can be equated to zero.

In many cases we use a free-body diagram and *force equilibrium equations* to determine the external forces and moments acting on a body or a part of a body. Once the critical sections are identified, free-body diagrams can be used to determine internal forces and moments, and then principles of strength of materials can be employed to calculate the stresses.

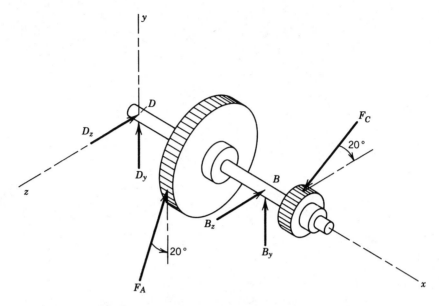

FIGURE 1-8. Free-body diagram of spur gear shaft.

The equations of force and moment equilibrium are

$$\Sigma F_x = 0 \quad \Sigma F_y = 0 \quad \Sigma F_z = 0$$

and

$$\Sigma M_{xx} = 0 \quad \Sigma M_{yy} = 0 \quad \Sigma M_{zz} = 0$$

Free-Body Diagram Analysis

For the system shown in Figure 1-9, the weight is suspended by a single cable that is fixed at A and passed over a small frictionless pulley at B. The pulley is suspended by a cable fixed at C. We can determine the force acting in the cable and the angle α.

First, we construct the free-body diagram of the idealized pulley as shown in Figure 1-10. From inspection of geometry

$$AB = 5 \text{ ft}$$

Force equilibrium requires that

$$F_{BC} \cos \alpha - F_{ABx} = 0 \tag{1-1}$$

$$F_{BC} \sin \alpha + F_{ABy} - 100 = 0 \tag{1-2}$$

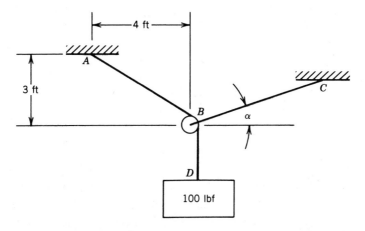

FIGURE 1-9. Cable pulley system supporting a weight.

where

$$F_{ABx} = 100(\tfrac{4}{5}) = 80 \text{ lbf}$$

$$F_{ABy} = 100(\tfrac{3}{5}) = 60 \text{ lbf}$$

Solving Eqs. (1-1) and (1-2) yields

$$\alpha = 26.6°$$

$$F_{BC} = 89.44 \text{ lbf}$$

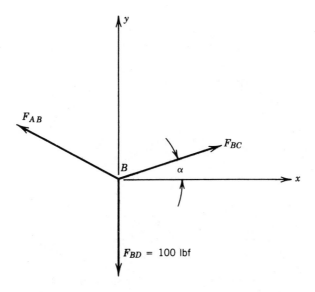

FIGURE 1-10. Free-body diagram of cable pulley system.

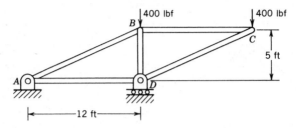

FIGURE 1-11. Pinned truss supporting two equal loads.

Figure 1-11 shows a truss whose members are connected at their ends by pin joints. The truss is hinged at A and simply supported at D. Neglecting the weight of the members, we can determine all the forces acting on each member of the truss for the given static loading. To determine these forces, we first construct a free-body diagram of the system as shown in Figure 1-12.

From inspection of the geometry

$$AB = CD = 13 \text{ ft}$$

Force equilibrium requires that

$$A_x = 0$$
$$A_y = 400 + 400 - D_y$$

$$(1\text{-}3)$$

If we sum the moments about point A to determine a second mathematical relationship between A_y and D_y, we have

$$12(D_y) - 12(400) - 24(400) = 0 \qquad (1\text{-}4)$$

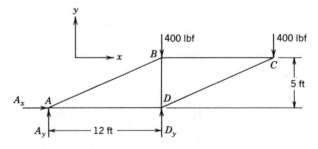

FIGURE 1-12. Free-body diagram of truss.

Solving Eq. (1-4) yields

$$D_y = 1200 \text{ lbf}$$

Using D_y in Eq. (1-3) gives

$$A_y = -400 \text{ lbf}$$

Now we construct the free-body diagrams for the isolated members as shown in Figure 1-13. Since the joints are pinned, only axial forces act on the members. If we apply the equations of force equilibrium to pin A, we have

$$F_{AD} + \left(\tfrac{12}{13}\right) F_{AB} = 0$$

$$\left(\tfrac{5}{13}\right) F_{AB} = 400$$

Solving these equations gives

$$F_{AB} = 1040 \text{ lbf}$$

$$F_{AD} = -960 \text{ lbf}$$

The remaining forces on members BC and BD can be determined by using a free-body diagram of pin B as shown in Figure 1-13.

From the equations of force equilibrium, we have

$$-\left(\tfrac{12}{13}\right) F_{AB} - F_{BC} = 0$$

$$-400 + F_{BD} - \left(\tfrac{5}{13}\right) F_{AB} = 0$$

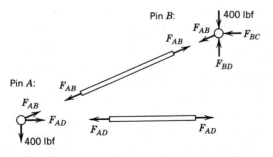

FIGURE 1-13. Free-body diagrams of individual truss members.

Since $F_{AB} = 1040$ lb, we can solve these two force equations to obtain

$$F_{BC} = -960 \text{ lbf}$$

$$F_{BD} = 800 \text{ lbf}$$

The minus sign in the answer for F_{BC} signifies that the initial direction assumed for this force in the free-body diagram was opposite to its true direction.

1-4 FORCE-FLOW CONCEPTS

The stress in a machine part is normally caused by the application of a force. For static equilibrium the force applied at one point is carried through the part and connected to some other structure that prevents acceleration of the part. This flow of force distributes itself as flux. The measurement of this force flux is equivalent to the measurement of the stress in the part at this point. Therefore methods that aid in visualizing force flux patterns can be used to identify stress distribution patterns.

The pattern of force flux lines can be visualized by association with heat flow or magnetic flux flow analogies. The point of force input can be regarded as a source and the reaction point can be designated as a sink. The principal requirements for all flux lines are that they originate at sources and terminate at sinks, that they do not intersect with each other, and that they remain within the body of the part. The actual path of these flux lines is determined by a set of partial differential equations. For first-order approximations; to avoid these complexities, application of the heat or magnetic flux flow analogy is justified.

Generally, flux lines seek to connect the source and sink points in the most direct manner. The flow of force through a part can be identified by flux lines, and the density of these flux lines indicates the intensity of the stress in the part. Discontinuities in part geometry can cause "bunching" or a concentration of flux lines. A greater density of flux lines indicates a higher stress. The part should be designed with gradual transitions between different sections to avoid these concentrations. The actual determination of flux-line patterns for a complex load and/or part is difficult. However, the specific stresses for designated points can be calculated by conventional means, then flux lines can be drawn between these points to gain insight into the overall stress pattern.

By being able to visualize the flow of stress through a part, we can identify the critical sections and stress concentrations. For example, with the rivet joint pictured in Figure 1-14, the flow of stress is sketched, and critical areas are identifiable.

The relative stiffness of various members in a machine part play a significant role in the flow of stress. Figure 1-15 shows a stiff member in parallel with a

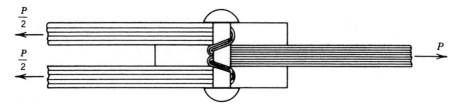

FIGURE 1-14. Force flow through a rivet joint.

flexible member. If member *A* is loaded in tension, member *B* carries very little load, and hence most of the force flows through member *A*.

The force-flow principle states that the flow of force distributes itself as a flux in a body and flows from the source of the applied load to the location of the reaction force on the body. Qualitative mental images of the loci of the force flux and the stress pattern are obtained.

Figure 1-16 shows the flow of flux in members with holes, notches, and steps. Critical areas are identifiable in locations of greatest flux-line density. Figure 1-17 illustrates the flow of force in a thermoplastic tabletop chain. The force flows from link to link through the connecting pin. A significant portion of the flow in the link goes through the center rib.

Figure 1-18 shows the force flow for three nut and bolt pairs. The force flows from the shank of the bolt through the threads and into the head of the nut. Concentrations of flux flow lines are seen, and because of the differences in geometry they are found in different locations for each case.

Diagrams that show the force flux lines for the case of a body that is accelerating could be drawn by using the same strategy as that for a static case, except that the effects due to the inertia forces should be considered; that is, the flow of inertia forces should be shown together with the flow of the normal loads.

The flow of force in assemblies of parts can be studied. Figure 1-19 shows a yoke joint loaded in tension; Figure 1-20 shows the force flow in the joint members; and Figure 1-21 shows in great detail the flow of force through a triple-riveted butt joint.

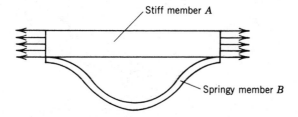

FIGURE 1-15. Force flow through stiff- and spring-like bars in parallel.

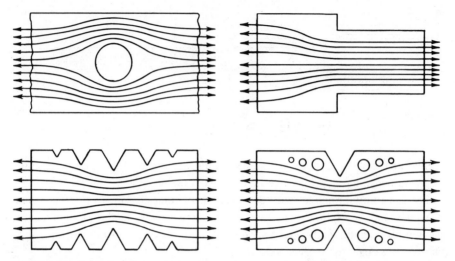

FIGURE 1-16. Force flow through members with steps, notches, and holes. [From Robert C. Juvinall, *Engineering Considerations of Stress, Strain and Strength*, McGraw-Hill, New York, 1967, p. 249. Reproduced by permission.]

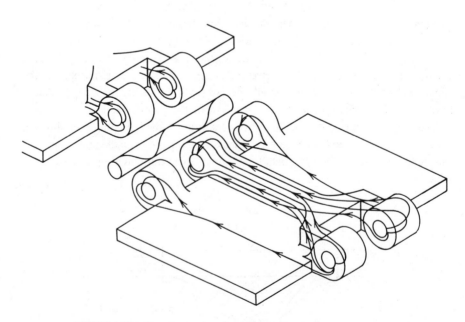

FIGURE 1-17. Flow through a thermoplastic tabletop chain link.

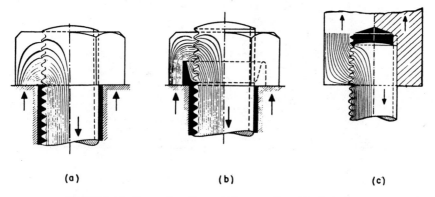

(a)   **(b)**   **(c)**

FIGURE 1-18. Force flow through three nut and bolt designs.

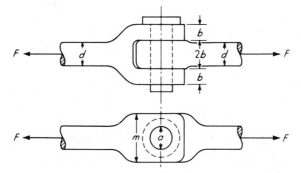

FIGURE 1-19. Yoke joint loaded in tension. [From Robert C. Juvinall, *Engineering Considerations of Stress, Strain, and Strength*, McGraw-Hill, New York, 1967, p. 12. Reproduced by permission.]

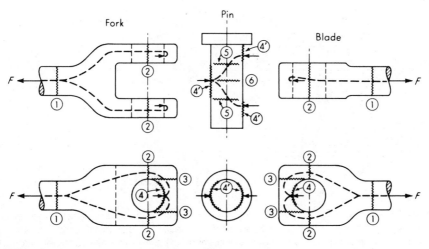

FIGURE 1-20. Flow of force and critical sections in yoke joint. [From Robert C. Juvinall, *Engineering Considerations of Stress, Strain, and Strength*, McGraw-Hill, New York, 1967, p. 12. Reproduced by permission.]

15

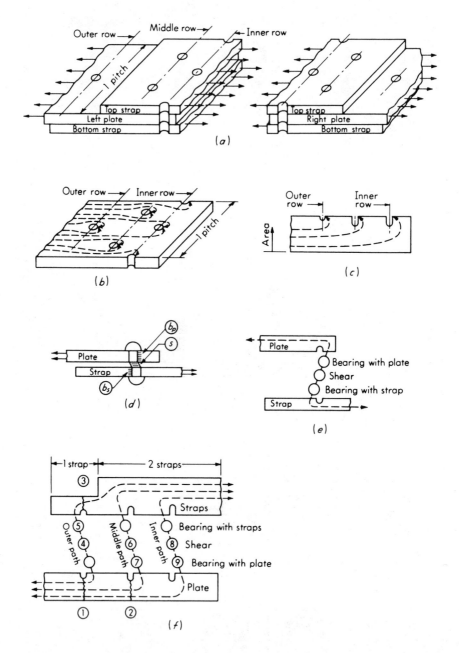

FIGURE 1-21. Flow of force through a triple-riveted butt joint. (*a*) Free-body diagram of sectioned joint, (*b*) force flow through plate to rivets, (*c*) diagram of force flow versus cross-sectional area, (*d*) force flow through rivet, (*e*) diagrammatic representation of force flow through rivet, (*f*) complete diagrammatic representation of force flow [From Robert C. Juvinall, *Engineering Considerations of Stress, Strain, and Strength*, McGraw-Hill, New York, 1967, p. 14. Reproduced by permission.]

1-5 DEFINITION OF NORMAL AND SHEAR STRESS

The *normal stress* over a given cross section at a certain point in a body is equal to the force per unit area acting perpendicular to the cross section at that point. If we consider a section through A-A in the system shown in Figure 1-22, define the cross-sectional area of the plate as A, and assume that the normal stress is uniformly distributed over the section A-A, then the normal stress at section A-A is

$$\sigma_A = \frac{P}{A}$$

Similarly, the normal stress at section C-C is

$$\sigma_C = \frac{(P/2)}{A} = \frac{P}{2A} = \frac{\sigma_A}{2}$$

Since the normal load carried by section C-C is half the load at section A-A (cross-sectional areas being equal), the normal stress at section C-C is half the stress at section A-A.

The *shear stress* over a given cross section at a certain point in a body is equal to the force per unit area acting tangential to the cross section at that point. If we consider a section through the rivet at B-B in the riveted joint, define the cross-sectional area of the rivet as a, and assume that the shear stress is uniformly

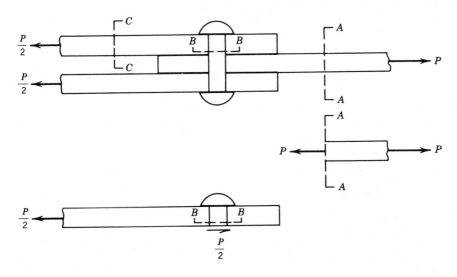

FIGURE 1-22. Force flow through a rivet joint with free-body diagrams showing shear and normal stresses.

distributed over the rivet cross section, then the shear stress at section *B-B* is given by

$$\tau_B = \frac{(P/2)}{a} = \frac{P}{2a}$$

1-6 CLASSIFICATION OF FORCES AND STRESSES

Stresses can basically be divided into the two classes, surface stresses and body stresses. *Surface stresses* are stresses that occur on the surface of a body such as in *spot contact* and bearing. Figure 1-23 shows spot contact between two crossed cylindrical bodies. Contact takes place at a single point or over a very small area. When contact between two bodies occurs over a relatively large area, as shown in Figure 1-24, the contact stress between the two bodies is called a *bearing stress*.

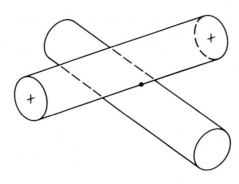

FIGURE 1-23. Spot contact between two crossed cylinders.

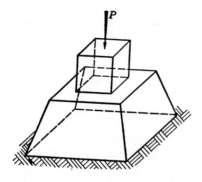

FIGURE 1-24. Bearing stress between blocks. [From Ref. 7, p. 9.]

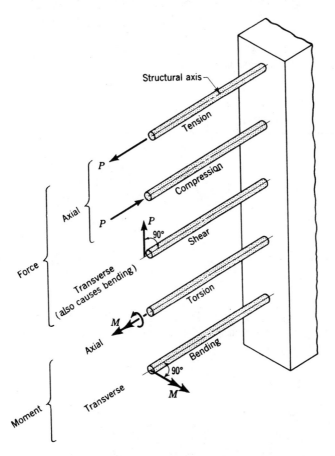

FIGURE 1-25. Classification of force transmission and corresponding stress. [From Ref. 31, p. 14.]

Body stresses are stresses that occur inside a body; that is, body stresses occur on internal cross sections of the body. Body forces and stresses can be categorized as shown in Figure 1-25. In the body part regions, removed from joints, there are several frequently encountered stress patterns: (1) bending, (2) tension and compression, (3) torsion, (4) transverse shear, (5) uniform shear, (6) contact, and (7) bearing. Figure 1-26 gives the shape of stress patterns found in body parts.

The primary parameters that determine the nature of the induced stress patterns are the *force system* and the *shape*. For example, in Figures 1-27a, b, and c, the shape is identical but the induced stress pattern is different in each case. In Figures 1-27a, and d, the system of forces are identical, yet because of a difference in shape the stress patterns differ.

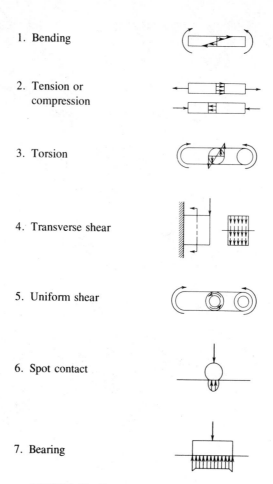

1. Bending

2. Tension or compression

3. Torsion

4. Transverse shear

5. Uniform shear

6. Spot contact

7. Bearing

FIGURE 1-26. Stress patterns in body parts.

Figure	Shape and Force System	Stress Pattern
(a)		Tension
(b)		Bending and transverse shear
(c)		Bending
(d)		Bending and transverse shear

FIGURE 1-27. Primary parameters that determine the stress pattern.

1-7 RELATIVE STRENGTHS OF STRESS PATTERNS

An intuitive feeling for the relative strengths of the various stress patterns can be obtained by calculating the stresses in several body parts, such as in a round bar in axial tension, in cantilever bending, and in torsion; in two crossed round bars loaded in compressive contact; and in a round bar loaded in bearing. Figure 1-28 pictures a variety of shapes and force systems for which the corresponding stress pattern and maximum stress are given.

Case	Description	Shape and Force System	Stress Pattern	Stress Formula	Stress (psi)
(a)	Round bar loaded in tension	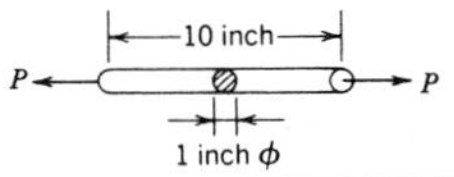	Tension	$\sigma = \dfrac{P}{A}$	$\sigma = 1,270$
(b)	Cantilever beam	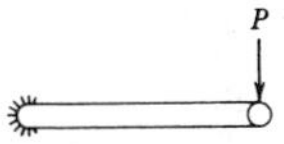	Bending	$\sigma = \dfrac{Mc}{I}$	$\sigma = 102,000$
(c)	Transverse shear stress	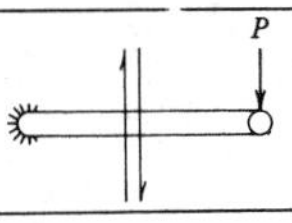	Transverse shear	$\tau = \dfrac{3P}{2A}$	$\tau = 1,700$
(d)	Torsion of solid bar	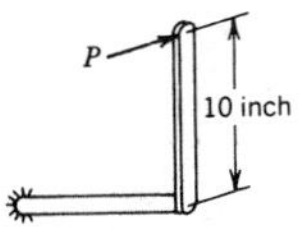	Torsion shear	$\tau = \dfrac{Tr}{J}$	$\tau = 51,000$
(e)	Torsion of hollow tube	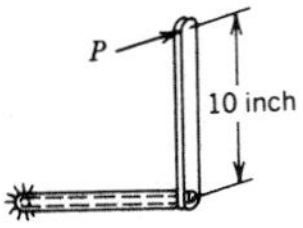	Uniform torsion shear	$\tau = \dfrac{Tr}{J}$	$\tau = 2,750$
(f)	Contact stress of 90° crossed cylinders	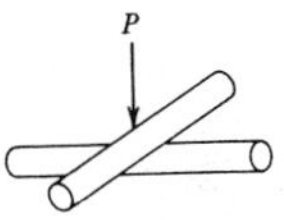	Contact		$\sigma = 640,000$
(g)	Bearing stress in clevis joint	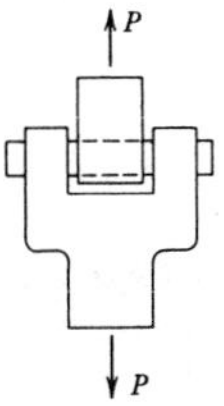	Bearing	$\sigma = \dfrac{P}{A}$	$\sigma = 1,000$

FIGURE 1-28. Maximum stress values for various stress patterns.

Case (a) Round Bar in Tension

The stress in the 1-in diameter round bar 10 in long, subject to a 1000 lbf in tension can be calculated by using

$$\sigma = \frac{P}{A}$$

where

$$P = 1000 \text{ lbf}$$

and

$$A = \frac{\pi d^2}{4} = \frac{\pi}{4}$$

Case (b) Round Cantilever Bar with End Load (Bending)

When the bar is arranged as a cantilever beam, the stress is a maximum at the wall where the bending moment M is largest. The moment

$$M = PL = (1000 \text{ lbf}) (10 \text{ in}) = 10,000 \text{ in-lbf}$$

The maximum bending stress is calculated as

$$\sigma = \frac{Mc}{I}$$

where

$$c = \frac{d}{2} = 0.50$$

and

$$I = \frac{\pi d^4}{64} = \frac{\pi}{64}$$

Case (c) Round Cantilever Bar with End Load (Shear)

The maximum transverse-shear stress for a round bar is calculated with

$$\tau = \frac{4V}{3A}$$

where for this case the transverse-shear force is

$$V = P$$

The area is

$$A = \frac{\pi}{4}$$

Case (d) Solid, Round Bar in Torsion
The 1000-lbf load is displaced 10 in from the axis of the cantilever beam and acts
in a plane perpendicular to this axis and at right angles to a radius drawn to the
force from the beam axis. Here the torsional shear stress is

$$\tau = \frac{Tr}{J}$$

where

$$J = \frac{\pi d^4}{32} = \frac{\pi}{32}$$

and

$$r = \frac{d}{2} = 0.50$$

The torque is

$$T = (1000 \text{ lbf})(10 \text{ in}) = 10{,}000 \text{ in-lbf}$$

Case (e) Hollow Round Bar in Torsion
The force system of case (d) is applied to a hollow tube of 5-in diameter having
the same cross-sectional area as the 1-in diameter solid bar. The torsional-shear
stress is again computed with

$$\tau = \frac{Tr}{J}$$

but for this case

$$J = \frac{\pi(d_0^4 - d_i^4)}{32}$$

Case (f) Crossed Solid Round Bars in Contact

The 1-in diameter bars have axes that are 90° askew. The bars are pressed together by a 1000-lbf load. The compressive contact stress can be calculated by using the equations of Hertz [4].

Case (g) Eye-and-Clevis Joint

The equation

$$\sigma = \frac{P}{A}$$

where A is the 1×1 in^2 projected area of the pin, is used to calculate a compressive stress that would result if the 1-in diameter solid bar were the pin in an eye-and-clevis joint subjected to a 1000-lbf pull.

1-8 NATURE OF EFFICIENT STRESS PATTERNS

Based on the material presented in this chapter and specifically on the information in Figure 1-28, we can make a list of stress patterns that seem to be efficient, and another list of those that appear to be inefficient. These lists are given in Table 1-2. The stress patterns that appear to be efficient, in general, are uniform, whereas the inefficient stress patterns are nonuniform.

1-9 ADDITIONAL COMPARISON OF STRESS PATTERNS

The relative strengths of various stress patterns can be studied further by determining the geometry where several stress patterns produce equal failure-inducing

TABLE 1-2. Efficient and Inefficient Stress Patterns

	Stress Pattern	Characteristic
Efficient	1. Tension 2. Compression 3. Uniform shear 4. Bearing	Stress is uniform
Inefficient	1. Bending 2. Torsion 3. Transverse shear 4. Spot contact	Stress pattern varies

Case	Description	Require	
(a)	Round bar cantilever loaded	Bending stress and the transverse-shear stress of equal failure-inducing intensity	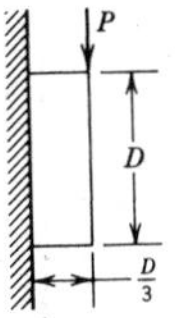
(b)	Eccentrically loaded round bar cantilever beam	Torsional stress equal to the transverse-shear stress	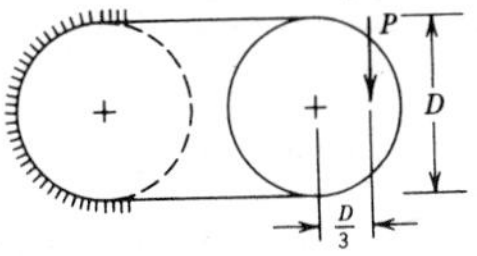
(c)	Eccentrically loaded round bar in tension	Bending stress equal to the tensile stress	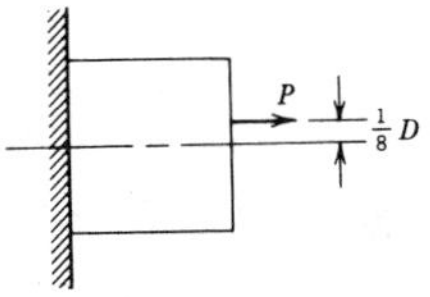

FIGURE 1-29. Geometries where stresses produce equal failure-inducing intensity.

intensity. Figure 1-29a gives the geometry for a round cantilever beam where the bending stress and the transverse-shear stress are of equal failure-inducing intensity. The geometry shown can be deduced by requiring that

$$\frac{\sigma}{2} = \frac{Mc}{2I} = \frac{3V}{2A} = \tau$$

Thus the length L must equal $D/3$. Figure 1-29b shows the geometry for an eccentrically loaded cantilever beam where the induced torsional stress is equal to the transverse-shear stress. The eccentricity, e, can be deduced by requiring that

$$\tau = \frac{Tr}{J} = \frac{3V}{2A}$$

Since $T = Pe$, the eccentricity is equal to $D/3$. Figure 1-29c gives the geometry for a round bar where a tensile force is applied offset to the bar axis so as to induce a bending stress that is equal to the tensile stress. For the condition where the bending stress equals the tensile stress, or

$$\sigma_{bending} = \frac{Mc}{I} = \sigma_{tensile} = \frac{P}{A}$$

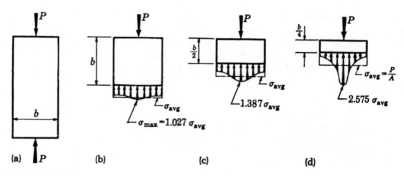

FIGURE 1-30. Stress distribution near a concentrated force [From Ref. 7, p. 42.]

the offset

$$e = \frac{1}{8} D$$

1-10 ST. VENANT'S PRINCIPLE

St. Venant's principle is related to how a concentrated surface stress (e.g., a stress due to a concentrated load) rapidly transforms into an evenly distributed body stress with depth of penetration into the part.

If a concentrated load P is acting on a body as shown in Figure 1-30, then the variation of the normal stress distribution is as given in the adjoining stress distribution diagrams. At a section close to the applied load, the peak of the normal stress will be relatively high. This nonuniform stress distribution rapidly smoothes out to a nearly uniform stress at a section below the surface equal to the width of the bar. St. Venant's principle states that the effect of forces or stresses applied over a small area may be replaced by a statically equivalent system that at a distance approximately equal to the width or thickness of a body causes a nearly uniform stress distribution.

2

EFFICIENT AND INEFFICIENT STRESS PATTERNS

In many cases, machine parts can be designed to avoid inefficient stress patterns of bending, torsion, transverse shear, and spot contact. The designer should know how to prevent these stress patterns or to modify the part shape or load system in order to obtain efficient patterns. Tension, compression, bearing, and uniform shear are efficient stress patterns. Bending, torsion, and spot contact are inefficient stress patterns.

2-1 BENDING STRESS

Figure 2-1 shows a dual beam system where two beams are subjected to bending. The top beam is subjected to a bending moment that varies from zero at each end to a maximum at the beam center. The midspan portion of the bottom beam is subjected to a constant bending moment. In both beams the bending stress varies from zero at the neutral surface (the center horizontal plane of the beam) to a maximum at the outermost fiber at each cross section of the beam. Since most of the material is not at maximum stress, the pattern is inefficient.

A typical bending stress pattern is pictured in Figure 2-2. Since a member in bending is not uniformly stressed throughout, bending is said to be an inefficient stress pattern.

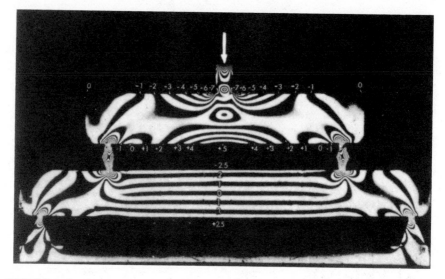

FIGURE 2-1. Isochromatic patterns for one beam loaded at three points and another beam loaded at four points. [From Ref. 1, p. 180.]

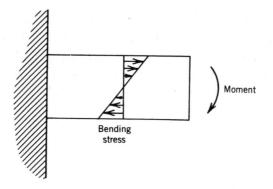

FIGURE 2-2. Stress distribution at a section of the beam calculated from the elementary theory of bending. [From Ref. 2, p. 2.6-2.]

2-2 TRANSVERSE-SHEAR STRESS

The transverse-shear stress distribution that develops when a moment varies along the length of a beam is shown in Figure 2-3. Transverse-shear stress is also an inefficient stress. Note that it is a maximum at the neutral axis and zero at the extreme outer surface of the beam.

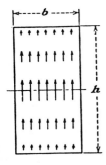

FIGURE 2-3. Shear-stress distribution across a beam of rectangular cross section. [From Ref. 3, p. 44.]

2-3 SPOT CONTACT STRESS

Spot contact is another inefficient stress pattern. Figure 2-4 shows the stresses in a gear tooth loaded in contact with another gear tooth. Inspection reveals large stresses at the point of contact and also large stresses at the gear tooth fillet where the bending stress is highest. The hook, shown in Figure 2-5, shows another case of spot contact and bending. A stress distribution for contacting cylinders of radius R_1 and R_2 is pictured in Figure 2-6.

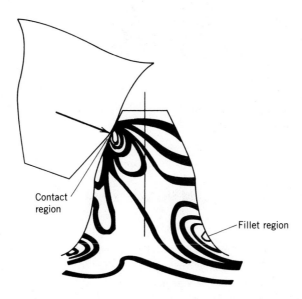

FIGURE 2-4. Isochromatic pattern in a loaded gear tooth showing high-contact stresses and bending stresses.

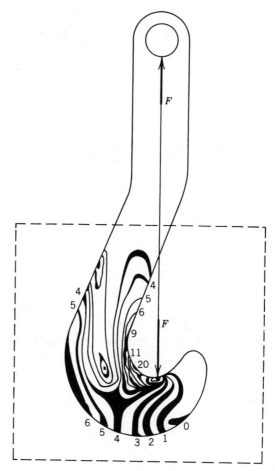

FIGURE 2-5. Isochromatic pattern of a loaded hook showing spot contact and bending stresses.

Figure 2-7 is an additional example of spot or Hertzian contact stress. This picture shows a disc loaded at opposite edges.

2-4 TORSION AND UNIFORM-SHEAR STRESS

A solid shaft subjected to torsion produces a stress distribution at each cross section that has a zero value in the shaft center and a maximum value at the outside diameter of the shaft. Thus the center portion of the shaft is relatively unstressed when compared to the outer portion of the shaft.

Figure 2-8 shows a hollow shaft subjected to torsion and the resulting shear-stress pattern that is closer to uniform than in a solid shaft. The hollow shaft in torsion gives a more efficient stress pattern.

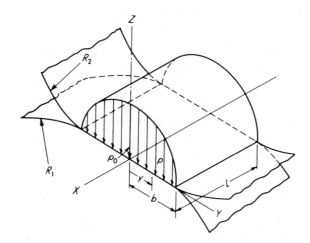

FIGURE 2-6. Contact pressure distribution for two cylinders with axes parallel. [From Robert C. Juvinall, *Engineering Considerations of Stress, Strain, and Strength*, McGraw-Hill, New York, 1967, p. 374. Reproduced by permission.]

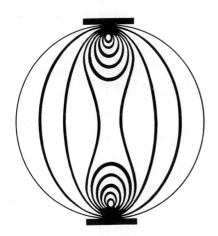

FIGURE 2-7. Fringe pattern for a disc loaded in compression.

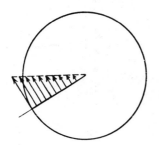 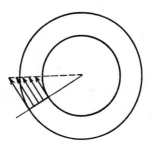

FIGURE 2-8. Stress distribution in a solid shaft and in a concentric hollow shaft both in torsion. [From Ref. 3, p. 24.]

2-5 USE OF INEFFICIENT STRESS PATTERNS

Inefficient stress patterns of bending, torsion, and spot contact are frequently used in good designs, even in those requiring high-performance capabilities. Rather than using uniform-stress patterns of tension, compression, and uniform shear, we employ inefficient patterns because of (1) space constraints, (2) motion requirements, (3) cost, (4) appearance, (5) conformity to manufacturing process, and (6) the necessity to use fasteners. Machine frames are generally designed within a constraint on the space required. Gears are used to produce a constant, angular speed ratio, but this requirement results in teeth that experience spot contact and are loaded in bending. Axles of vehicles employ solid rather than hollow round bars because of cost considerations. Chairs have cantilevered arms because of an improved appearance. Machines have joints because of the difficulty of fabricating a machine as one monolith. Fasteners are used to bolt, spline, and hinge parts together even though they produce inefficient stress patterns.

Although uniform-stress patterns are desired for a high strength-to-weight ratio, there are numerous cases where inefficient stress patterns are employed because of appearance (wraparound windshield) or performance requirements (gears).

Figure 2-9 shows a maple syrup valve. Bending stresses exist in the cantilever portions of the valve; spot contact stresses are found at the interface between the ball and the cantilever. Yet these inefficient patterns arise because of space and functional constraints that are required for proper valve operation.

2-6 SHAPE CHANGES TO OBTAIN UNIFORM STRESS

Figure 2-10 pictures a C-clamp loaded with two opposing forces. The width of the beam increases as the distance from the point of application of the load increases.

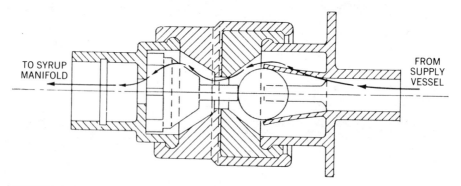

TO SYRUP MANIFOLD

FROM SUPPLY VESSEL

FIGURE 2-9. Maple syrup valve where integral cantilever springs retain the stem portion in the flow position unless there is backflow. [From Ref. 6, p. 120.]

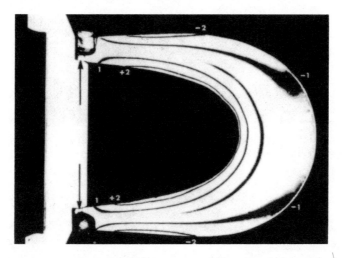

FIGURE 2-10. Isochromatic pattern in the curved beam portion of a press designed for constant tensile stress at the inner edge. [From Ref. 1, p. 185.]

The increase in width results in the increase in moment being balanced by an increase in moment of inertia. The net effect is a fairly uniform-stress distribution, with the maximum bending stress remaining approximately the same on successive sections of the C-clamp.

A picture of a beam, fat in the middle, is shown in Figure 2-11. Because the beam has an increasing depth, the outermost fibers along the entire length of the beam have the same maximum stress. If the beam had a constant uniform cross section, the maximum bending stress would be found at the outermost fiber at the midspan of the beam since the bending moment is maximum at the beam center.

Figure 2-12 shows an isochromatic pattern for a tapered beam with central load. The stress at the extreme fibers is indeed fairly uniform with fringe numbers varying from 4 to 6. The fringe number is an indicator of the shear stress. In general, the higher the fringe number is, the higher the shear stress is. The parabolically shaped

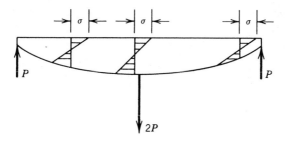

FIGURE 2-11. Stress patterns for a variable depth beam designed for constant outer fiber stress along its entire length. [From Ref. 2, p. 3.1-6.]

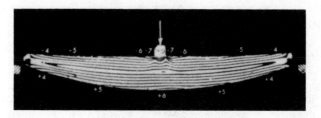

FIGURE 2-12. Isochromatic pattern in a beam of parabolically varying depth under three-point loading. [From Ref. 1, p. 184.]

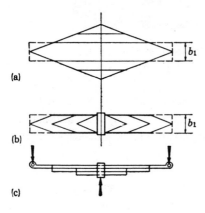

FIGURE 2-13. Leaf spring assembled from a beam of constant strength. [From E. P. Popov, *Mechanics of Material*, © 1952, p. 398. Reprinted by permission of Prentice-Hall, Inc., Englewood Cliffs, N.J.]

beam would produce a maximum stress that is approximately constant for *all positions* of the vertical load. A leaf spring is another common type of tapered beam (see Fig. 2-13).

3

DESIGNING
FOR RIGIDITY

For many machine components, rigidity requirements must be met to ensure that linear or angular deflection is kept small so as not to affect machine performance.

Components are made rigid to minimize misalignment; avoid interference; and reduce noise, stress, and wear rates. Shaft deflection, for example, can cause shaft and bearing bore misalignment, cause interference between adjacent components assembled on the shaft, and cause misalignment in attached gears or sprockets and chains. Misalignments and interferences result in increased noise, wear rates, and stresses and can change system operating characteristics. Roller element bearing misalignment caused by shaft deflection can severely decrease the life of the bearing.

There are machine elements where rigidity is unfavorable and flexibility is desired, for example, a degree of flexibility is required in snap rings or retaining rings for assembly. Yet a degree of rigidity is required to maintain the snap ring in position. Flexible bolts are desired in preloaded bolted joints and for bolts required to carry higher impact loads, as these bolts provide a better load distribution and a higher load capacity for the joint itself. Flexible teeth on gears and sprockets and flexible rims generally provide a better load distribution with mating machine parts and hence provide a longer life. Springs are also required to be flexible as large elastic deformations or deflections are required.

Rigidity can be defined as the amount of deflection per unit of applied external load. Rigidity of a machine component results from both the volume deflection as well as the surface deformation. The modulus of elasticity of a material is an indication of the material's rigidity.

35

3-1 SPRING CONSTANT

The *spring constant* is a measure of a machine part's rigidity. We consider a simple part comprised of a bar of length L having a uniform cross section and made up of a material that obeys Hooke's law. For an axial force applied, where the force does not stress the material above the elastic range, the spring constant is given as

$$k = \frac{F}{y} = \frac{AE}{L}$$

where

k = spring constant

F = applied force

y = deflection of one end of the bar relative to the other

A = cross-sectional area

E = modulus of elasticity

L = length of the bar

For the same bar loaded in pure torsion

$$k_t = \frac{T}{\theta} = \frac{JG}{L}$$

where

k_t = torsional spring constant

T = torque applied

θ = angular rotation of one end relative to the other

J = polar moment of inertia of the cross section

G = modulus of torsional rigidity

L = length of the bar

Spring constants for members in parallel can be computed by using

$$k = \sum_{i=1}^{n} k_i$$

or for elements in series

$$k = \frac{1}{\sum_{i=1}^{n} (1/k_i)}$$

Adding a stiff element in parallel with a very flexible element increases the overall stiffness, whereas adding a stiff element in series with a flexible one has a negligible effect on increasing stiffness.

The deflections for beam-like machine elements such as shaft axles, levers, cranks, brackets, springs, and wheels can be determined by treating these elements as beams.

Roark [30] provides calculations, giving slopes and deflections for numerous beams and frames having uniform sections with various applied loads. Appendix 7 gives formulas for calculating deflections for several beams. Spring constants and deflections for more complex elements such as stepped shafts can be computed by solving the equation

$$\frac{d^2y}{dx^2} = \frac{M}{EI} \qquad (3\text{-}1)$$

where

M = moment along the shaft

E = modulus of elasticity

I = moment of inertia of the cross section

y = deflection of the shaft

x = distance along the shaft

Rules of thumb for rigidity of various machine components are available for certain machine elements. For example, in frames the limit of deflection is $1/360$ of the span of the frame member. For shafts, the maximum deflection is held to $1/100$ of the shaft length.

3-2 RELATIVE RIGIDITY OF STRESS PATTERNS (OR LOADING SYSTEMS)

Machine component rigidity depends on the material, component shape, force system type, and directions of the applied loads.

An intuitive feeling for the relative rigidity of the various stress patterns and loading systems can be obtained by calculating the stiffnesses of several body parts, such as a round bar in axial tension, in cantilever bending, in transverse shear, and in torsion; a hollow shaft in torsion; two crossed round bars loaded in compression contact; and a round bar loaded in bearing. Figure 3-1 pictures a variety of shapes and force systems for which the corresponding stress pattern, stiffness formula, and stiffness are given.

Case (a) Round Bar in Tension
The stiffness, k, in the 1-in diameter round steel bar 10 in long, subject to a tension load, can be calculated by using

$$k = \frac{AE}{L} = \left(\frac{\pi}{4}\right)\frac{(30 \times 10^6)}{10} = 23.6 \times 10^5 \text{ lbf/in}$$

where

$A = \pi d^2/4 = $ cross-sectional area

$E = $ modulus of elasticity of steel

$L = $ length of the bar

The term AE is referred to as the *tensile* (or compressive) *rigidity* of the section.

Case (b) Round Cantilever Bar with End Load (Bending Stiffness)
The stiffness of the preceding bar when arranged as a cantilever beam, but considering only the bending deflection and neglecting the deflection due to shear, is given by

$$k = \frac{3EI}{L^3} = \frac{[(3)(30 \times 10^6)(\pi/64)]}{10^3} = 4.4 \times 10^3 \text{ lbf/in}$$

The term EI, which is the product of the modulus of elasticity and the second moment of area of the cross section about the neutral axis, is usually referred to as the *flexural rigidity* and is used in many bending deflection formulas.

Case (c) Round Cantilever Bar with End Load (Shear Stiffness)
The stiffness of the bar arranged as a cantilever beam, but considering only the shear deflection and neglecting the deflection due to bending, is computed by using

$$k = \frac{9AG}{10L} = \frac{[9(\pi/4)(11.5 \times 10^6)]}{(10)(10)} = 81.3 \times 10^4 \text{ lbf/in}$$

The term AG is referred to as the *transverse rigidity* of the section.

Case (d) Solid, Round Bar in Torsion

The torsional stiffness, k_t, of the round cantilever bar is given by

$$k_t = \frac{JG}{L} = \frac{[(\pi/32)(11.5 \times 10^6)]}{10}$$

$$= 1.13 \times 10^5 \text{ lbf-in}$$

where

$J = \pi d^4/32 = $ polar moment of inertia

$G = $ torsional modulus of steel

$L = $ length of the bar

The term GJ, which is the product of the modulus of rigidity and the polar moment of inertia is usually referred to as the *torsional rigidity* of the section and always appears in torsional deflection computations.

Case (e) Hollow Round Bar in Torsion

If a steel cantilever bar is hollow as shown in Figure 3-1e, and has an outside diameter $d_o = 5$ in and an inside diameter $d_i = 4.9$ in such that the area of the cross section is the same as the area for the solid, 1-in diameter bar, the torsional stiffness of the hollow bar is

$$k_t = \frac{JG}{L} = \frac{[(\pi/32)(d_o^4 - d_i^4)(11.5 \times 10^6)]}{10}$$

$$= 55.3 \times 10^5 \text{ lbf-in}$$

where

$$J = \left[\frac{\pi(d_o^4 - d_i^4)}{32} \right]$$

Case (f) Crossed Solid Round Bars in Contact

Consider two 1-in diameter crossed bars pressed together by a force that is perpendicular to the axes of the crossed bars. If the bars are steel and have a yield strength of 180,000 psi (pounds per square inch), a Hertzian stress analysis will show that yielding in the contact region will begin where the applied force $P = 27.6$ lbf.

The stiffness of contact for the two cylinders at the load where the material is about to undergo plastic deformation can be calculated following the methods given

Case	Description	Shape and Force System	Stress Pattern	Stiffness Formula	Stiffness
(a)	Round bar loaded in tension		Tension	$\dfrac{AE}{L}$	23.562×10^5 (lbf/in)
(b)	Cantilever beam		Bending	$\dfrac{3EI}{L^3}$	$.0442 \times 10^5$ (lbf/in)
(c)	Transverse-shear stress		Transverse-shear stress	$\dfrac{9AG}{10L}$	8.1289×10^5 (lbf/in)
(d)	Torsion of a solid bar of 1-in ϕ		Torsional-shear stress	$\dfrac{JG}{L}$, where $J = \dfrac{\pi d^4}{32}$	1.129×10^5 (lbf/in)

(e)	Torsion of a hollow tube of 5-in outer diameter and the same cross-sectional area as case (d)	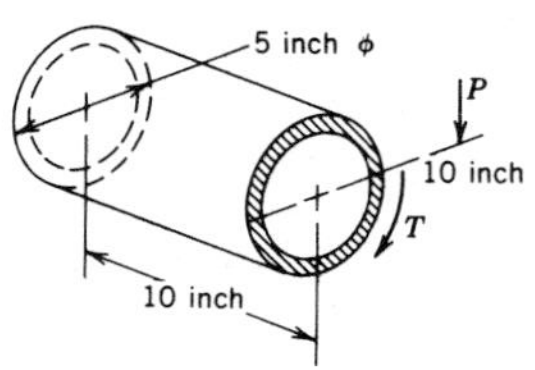 Torsional-shear stress	JG/L, where $J = \dfrac{\pi(d_o^4 - d_i^4)}{32}$	55.321×10^5 (lbf/in)
(f)	Two round bars pressed together with their axes at 90° to each other	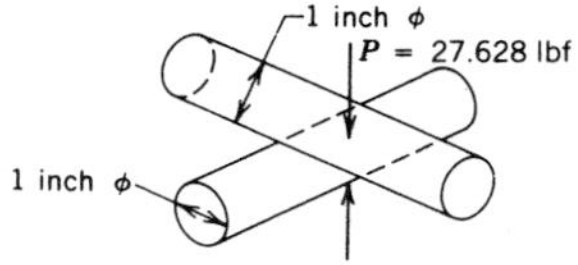 Hertz contact stress	Refer to Ref. 23 [pp. 516–517]	2.8256×10^5 (lbf/in)
(g)	Eye-and-clevis joint of 1-in length	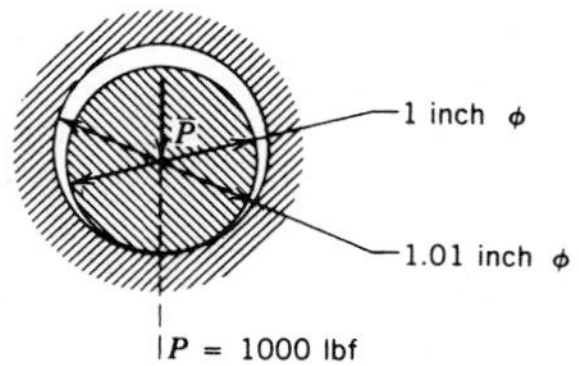 Hertz contact stress (conformal contact stress)	Refer to Ref. 23 [pp. 516–517]	83.285×10^5 (114.84×10^5) (lbf/in)

FIGURE 3-1. Comparison of the rigidities under different loading patterns (stress patterns).

in Ref. 23 [pp. 516–517]. The spring constant is

$$k = 9.3 \times 10^4 \, P^{1/3} \, \text{lbf/in}$$

When $P = 27.6$ lbf, $k = 2.8 \times 10^5$ lbf/in. The value for k was obtained by considering only the deformation due to Hertz contact stress. The additional deformation due to the elasticity of the body portion of the cylinders can be shown, using the analysis given in Ref. 24 [p. 493], to be negligible.

Case (g) Eye-and-Clevis Joint
We consider an eye-and-clevis joint subject to a 1000 lbf pull having a 1-in diameter solid pin with a clearance of 0.010 in in the joint as shown in Figure 3-1g. The length of the joint is 1 in. Assuming the contact is Hertzian and using the equations given in Ref. 23 [p. 517], we find that the effective stiffness of the eye-and-clevis joint is $k = 8.3 \times 10^6$ lb/in. The maximum contact stress is found to be 10,195 psi. This is about 10 times as high as the value given by $\sigma = P/A = 1000$ psi where $P = 1000$ lbf and $A = 1$ in^2, the projected area.

The relative stiffnesses of various geometries and loading systems can be studied further by determining the geometry where several stiffnesses are equal in magnitude. Figure 3-2 gives the geometry for a round cantilever beam. Suppose we are interested in determining the relationship between the beam length L and the diameter D such that

$$k_b = k_{ts} \tag{3-2}$$

where for bending, the spring constant is

$$k_b = \frac{3EI}{L^3} = \frac{3E[\pi D^4/64]}{L^3} = \frac{3\pi ED^4}{64L^3} \tag{3-3}$$

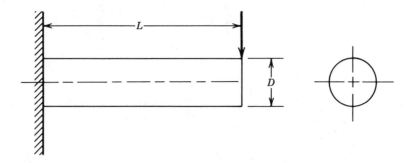

FIGURE 3-2. Cantilever beam.

and for transverse shear, the spring constant is

$$k_{ts} = \frac{9AG}{10L} = \left[\frac{9G}{10L}\right]\left[\frac{\pi D^2}{4}\right] = \frac{9\pi GD^2}{40L} \tag{3-4}$$

Combining these equations gives, for equal bending stiffness and transverse-shear stiffness

$$D = \left[\frac{5G}{24E}\right]^{1/2} L$$

or

$$L = \left[\frac{5E}{24G}\right]^{1/2} D$$

For a steel bar with $E = 30 \times 10^6$ psi and $G = 11.5 \times 10^6$ psi

$$\frac{L}{D} = 0.74$$

Figure 3-3 shows a round bar that is eccentrically loaded. We are interested in determining a relationship between L_1 and L such that the stiffness (deflection) at the point of loading is the same due to both torsion as well as bending.

The vertical deflection at the point of loading due to bending is

$$y_b = \frac{PL^3}{3EI} = \frac{64PL^3}{3\pi ED^4}$$

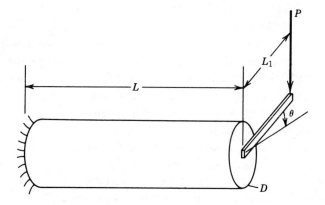

FIGURE 3-3. Eccentrically loaded round bar.

The rotation of the bar due to the torque PL_1 is

$$\theta = \left(\frac{PL_1}{J}\right)\left(\frac{L}{G}\right) = \frac{32PLL_1}{\pi GD^4}$$

The vertical deflection at the point of loading due to torsion is

$$y_t = \theta L_1 = \frac{32PLL_1^2}{\pi GD^4}$$

For

$$y_b = y_t$$

we require

$$L_1 = L\left[\frac{(2G)}{3E}\right]^{1/2}$$

For steel with $E = 30 \times 10^6$ psi and $G = 11.5 \times 10^6$ psi

$$L_1 = 0.51L$$

3-3 RIGIDITY OF VARIOUS STRESS PATTERNS

Inspection of Figure 3-1 reveals that the tensile stress case is more rigid than the bending, transverse-shear, and spot contact (or line contact) cases. The transverse-shear case, although not so rigid as the normal stress case, is considerably more rigid than the bending case.

By comparing cases (d) and (e), we can conclude that although both members have the same cross-sectional area, that is, the same volume of material, the hollow tube has about 50 times the torsional rigidity as the solid rod, because the hollow section is almost uniformly stressed throughout whereas the solid shaft is not. Thus it appears that uniform stress implies increased rigidity.

In case (f), where we studied Hertz contact stress for crossed cylinders, it is evident that plastic deformation takes place at a relatively low load of 27.6 lbf, whereas in the tension, bending, and transverse-shear cases yielding does not occur even under a load of 1000 lbf. This is explained by the extremely inefficient load-carrying capacity caused by the stress pattern of spot contact.

In case (g), where the clearance between the shaft and the hole is small and equal to 0.01 in, the type of contact taking place is of the conformal contact type and not of the spot contact type. Hence the material does not yield even at a load of 1000 lbf. Also, the effective stiffness in case (g) is much higher than that in case (f) because case (g) is conformal contact whereas case (f) is nonconformal contact.

3-4 NATURE OF RIGID BODIES

Higher rigidity is displayed by bodies under tensile or compressive stress as opposed to bending, transverse-shear, torsion, or spot contact stress. Bodies that are shaped such that the material is uniformly stressed throughout will be more rigid, as can be seen by comparing case (d) with case (e). A general principle that can serve as a guide to the form design of rigid body parts is *design for uniform stress*.

3-5 DESIGN RECOMMENDATIONS FOR INCREASING RIGIDITY

The rigidity of a machine element can be increased by (1) increasing the element cross section; for example, for a bar in tension, doubling the area would double the stiffness, and (2) making the element shorter in length; for example, for a bar in tension, the stiffness varies inversely as the length. For a member subject to torque, a decrease in the moment arm would increase the stiffness measured at the point of load application.

In summary, for increased rigidity

1. Increase the cross-sectional size or decrease the length.
2. Reduce moment arms.
3. Encourage changes to obtain uniform-stress patterns of tensile or compressive direct stress, uniform-transverse, torsion, or shear stress.

Uniform-stress patterns can be obtained by employing cross sections that resist bending, such as that of an I-beam, and for members subject to torsion, by employing a thin-walled tube. Also, more uniform-stress patterns can be obtained by adding supports that place members that are in bending into tension or compression. For example, the addition of an end support for a cantilever beam will create a triangular arrangement of members which generates uniform tension or compression in each member. This results in an increased volume rigidity.

Contact deformations are also important in component rigidity, for example, in cams, cam followers, gear teeth, and ball and roller bearings. Point and line contact does not give a rigid joint. Hertzian theory can be used to calculate contact stresses and deflections. More information on contact deformation calculations can be obtained by referring to Ref. 25 [pp. 372–392].

Contact rigidity can be improved by (1) reducing the number of contacting joints; (2) improving the surface quality of the contacting areas, providing improved surface finishes and better mating contours; (3) preloading to produce a stiffer contact joint, for example, in ball or roller element bearings where stiffness increases with load. This is also true for bolted connector joints where pretensioning results in a stiffer joint element; and (4) using a higher viscosity oil between contacting surfaces to make the contact more rigid. The use of an oil between nonlubricated surfaces will increase contact stiffness.

The rigidity of a machine is influenced by the joint stiffness. Joints that fasten machine components and assemblies together can be made stiffer by (1) employing, for a given joint, many more fasteners; (2) making these fasteners as short as possible especially where the joint is loaded in tension or compression; (3) designing the joint to obtain a uniform stress; and (4) employing parts to gather and gradually channel stress through the joint in a uniform manner.

4

GENERAL PRINCIPLE
FOR FORM DESIGN

Based on the discussions in the previous chapters, we can formulate a general principle that will serve as a guide to the form-design of the body portions of all machine parts where stresses are to be minimized or where greater strength and rigidity are required while using less material.

4-1 GENERAL PRINCIPLE

The general principle is *design to get uniform stress*. Corollaries to this principle include the (1) tetrahedron-triangle principle, (2) uniform-shear or hollow-tube (or shaft) principle, (3) forcing-load transfer principle, and (4) the mating-surface principle.

4-2 TETRAHEDRON-TRIANGLE PRINCIPLE

The use of tetrahedron and triangle shapes results in uniform stresses in tension or compression, and these stress patterns are the most efficient. Triangles are found in bicycle frames, light poles, and machine frames. Tetrahedrons are seen in television and radio towers, crane booms, and oil rig derricks.

4-3   UNIFORM SHEAR OR HOLLOW SHAFT PRINCIPLE

Uniform shear is obtained with a hollow shaft or tube, which allows for the transmission of greater torsional loads with less weight. Helicopter and auto gyro rotors use hollow shafts.

4-4   FORCING-LOAD PRINCIPLE

In many machine parts, certain portions of the body of the part exist for the purpose of forcing other portions to be loaded essentially in tension, compression, or uniform shear. For example, the vertical web of an I-beam forces the top and bottom flange portions to become loaded in tension and compression. Cross members in beams and webs in C-channels provide the same forcing purpose.

4-5   MATING-SURFACE PRINCIPLE

Matching the shapes of contacting surfaces of mating parts produces a more efficient transmission of force between the parts. For example, a round handle on a machine allows an improved match with the surface of a gripping hand. Contoured keyboard keys produce an improved match for the typist's fingers.

The corollaries to the general principle of designing to get uniform stress can be applied as rules to obtain stress patterns of tension, compression, uniform shear, and bearing. However, even machine parts with uniform stress throughout may often be further improved by selecting a configuration that optimizes the strength-to-weight ratio.

4-6   LOAD-LEVER PRINCIPLE

The load-lever principle recognizes that triangle and tetrahedron shapes can have a configuration that provides a maximum load-carrying capacity for a given weight. For example, in the structure shown in Figure 4-1, one value of the angle α will minimize the weight of the two diagonally oriented members, yet resist the same force F. In determining this value of α, we do not consider buckling; F, b, and the strength, S, are fixed, and the angle α is a variable.

The weight, W, of the members is proportional to their length. If we assume that the weight per unit length is 1, then

$$W = \frac{2b}{\cos \alpha}$$

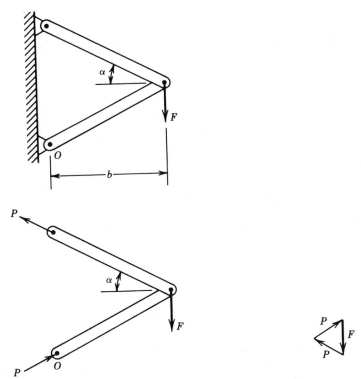

FIGURE 4-1. Triangular structure and corresponding free-body diagram and force triangle.

Summation of moments about point O gives

$$P = \frac{F}{(2 \tan \alpha \cos \alpha)} \sim S^{-1}$$

The weight-to-strength ratio

$$\frac{W}{P^{-1}} = \frac{2bF}{2 \tan \alpha \cos^2 \alpha} = \frac{bF}{\sin \alpha \cos \alpha}$$

We want W/P^{-1} a minimum, and since bF is a constant, we thus want the quantity $(\sin \alpha \cos \alpha)$ a maximum. Let

$$F(\alpha) = \sin \alpha \cos \alpha$$

Then

$$F'(\alpha) = \cos \alpha \cos \alpha - \sin \alpha \sin \alpha = \cos^2 \alpha - \sin^2 \alpha$$

Setting

$$F'(\alpha) = 0$$

yields

$$\cos^2 \alpha - \sin^2 \alpha = 1 - \sin^2 \alpha - \sin^2 \alpha = 1 - 2 \sin^2 \alpha = 0$$

or

$$\sin^2 \alpha = \tfrac{1}{2}$$

and

$$\alpha = \sin^{-1} \sqrt{\tfrac{1}{2}} = 45°$$

The load-lever principle may also be applied to more elaborate pin-jointed frames to achieve structures containing the least volume of material [22]. Figure 4-2 shows a symmetrically loaded symmetrical frame with the force $2F$ applied at node 5. A

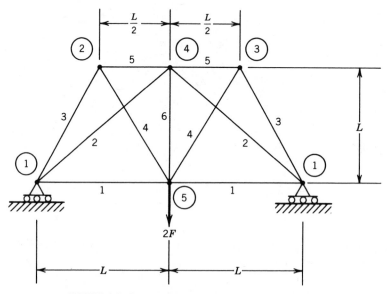

FIGURE 4-2. Symmetrical frame with central load.

maximum strength-to-weight design is desired and can be achieved by omitting members 2 and 6, and for member 1, having a cross-sectional area A_1, taking

$$A_3 = A_4 = \sqrt{\tfrac{5}{2}}\, A_1$$

and

$$A_5 = A_1$$

The pin at node 4 should also be eliminated to avoid instability. For details and background theory, see Ref. 22.

5

TENSION, COMPRESSION, AND BENDING

Tensile and compressive loads produce very efficient stress patterns in axially stressed members of machines and structures. The type of load, the shape, and the change in the cross-sectional area along the length of the element are factors that influence the uniformity of tensile- and compressive-stress patterns. An efficient stress pattern in bending is one where a large percentage of the total cross section is stressed to the allowable limit.

5-1 TENSION AND COMPRESSION

Consider the stepped bar shown in Figure 5-1. The average tensile stress at a given section will vary with the cross section, as shown by the spacing of the force-flow lines; that is, the force flux varies along the element. The *average* tensile (or compressive) stress

$$\sigma = \frac{P}{A} \tag{5-1}$$

where

P = load

A = cross-sectional area

52

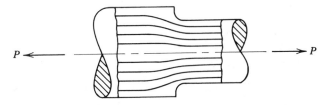

FIGURE 5-1. Force flux in a stepped bar.

Although the average tensile and compressive stress at a given section depends only on the load and the cross-sectional area, the localized stress will depend on other factors such as stress concentration and effect of concentrated loads acting at a point close to where the stress is computed (i.e., St. Venant's effect), all of which tend to increase the localized stress.

Efficient shapes for tension and compression are constant area shapes that are free from effects such as stress concentration and the action of concentrated loads.

5-2 BENDING

Bending is the result of the action of a couple or a transverse force on an element. The factors that influence the bending stress pattern are the loading and the shape:

1. *Loading.* Both the type of load and the magnitude of the load, will affect the bending moment at a given section.

2. *Shape.* a. The shape of the machine part affects the bending moment, which in turn affects the bending stress.

   b. The cross-sectional shape determines the moment of inertia of the given section that in turn determines the bending stresses.

Consider a section of a machine part or structural element under the action of a pure bending moment (no axial load). For the case of pure bending, the axis of zero stress and strain is called the *neutral axis*. An equation for the bending stress in the element can be developed assuming [34]:

1. The material is homogeneous, is isotropic, and has the same value of Young's modulus in tension and compression.

2. The beam is initially straight and all longitudinal filaments bend into circular arcs with a common center of curvature.

3. Transverse cross sections remain plane and perpendicular to the neutral surface after bending.

4. The radius of curvature is large compared with the dimensions of the cross section.

5. The stress is purely longitudinal and local effects near concentrated loads will be neglected.

Under these assumptions the bending stress is

$$\sigma = \frac{My}{I} \tag{5-2}$$

where

M = bending moment acting at the section where the bending stress is required

I = moment of inertia of the cross-sectional area about the neutral axis

y = distance from the neutral axis to the point where the bending stress is required (see Fig. 5-2)

We let the maximum value of

$$y = y_{max} = c$$

Then, the maximum bending stress at the given section is

$$\sigma_{max} = \frac{Mc}{I} = \frac{M}{(I/c)} \tag{5-3}$$

The ability of a given cross section to carry bending moments (i.e., the bending moment that can be carried for a given maximum stress) depends on the ratio (I/c). This ratio is commonly referred to as the *section modulus* of the cross section. The larger the value of (I/c) is, the lower is the maximum bending stress for a given bending moment. To improve the bending strength-to-weight ratio for

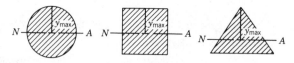

FIGURE 5-2. Maximum distances from the neutral axis for three shapes.

a member with a given cross-sectional area, the section modulus should be increased while keeping the cross-sectional area constant.

5-3   BENDING—STRENGTH-TO-WEIGHT RATIO

Equation (5-3) can be rewritten as

$$M = (\sigma_{\max}) \left(\frac{I}{c} \right)$$

For a given stress ($\sigma_{\max}$), the bending moment strength of a given section is directly proportional to the section modulus (I/c). Furthermore, the weight of the given element is directly proportional to its cross-sectional area. Hence the (section modulus)/(cross-sectional area) gives an accurate measure of the bending strength-to-weight ratio of a given section when compared to other sections having the same strength and density.

Table 5-1 lists equations for typical sectional properties and includes equations for the section modulus/area ratio. Table 5-2 shows dimensioned cross sections and lists values for the section modulus/area ratio which is proportional to the strength-to-weight ratio.

An efficient stress pattern in bending is obtained in a system where a high percentage of the total cross section is stressed to the maximum stress. Since the bending stress is proportional to the distance from the neutral axis, an efficient bending stress pattern can be achieved if the cross section is shaped such that a higher fraction of the cross-sectional area is placed away from the neutral axis. Indeed, beams that have high-bending strength will have most of their material placed away from the centroidal axis such as in *I*-sections and channel sections. This type of geometry produces a higher moment of inertia for a given depth of the cross section, thus maximizing the section modulus.

5-4   CONFIGURATIONS THAT ARE EFFICIENT FOR BENDING

Along with *I* and channel beams, composite and sandwich construction beams are sections that are efficient for bending.

Composite Sections

Structural elements under bending loads and manufactured from a relatively soft and weak material such as wood can be reinforced by stiffening with steel plates

TABLE 5-1. Typical Section Properties

Section	Section Area A	Moment of Inertia About x-x Axis I_x	Section Modulus About x-x Axis Z_x	Section Modulus/Section Area $\propto$ Strength/Weight
	$\dfrac{\pi d^2}{4}$	$\dfrac{\pi d^4}{64}$	$\dfrac{\pi d^3}{32}$	$\dfrac{d}{8}$
	$\pi(R_0^2 - R_i^2)$	$\dfrac{\pi}{4}(R_0^4 - R_i^4)$	$\dfrac{\pi(R_0^4 - R_i^4)}{4R_0}$	$\dfrac{(R_0^2 + R_i^2)}{4R_0}$
	$2\pi rt$	$\pi r^3 t$	$\pi r^2 t$	$\dfrac{r}{2}$
	bh	$\dfrac{bh^3}{12}$	$\dfrac{bh^2}{6}$	$\dfrac{h}{6}$
	$bh - b_0 h_0$	$\dfrac{bh^3 - b_0 h_0^3}{12}$	$\dfrac{bh^3 - b_0 h_0^3}{6h}$	$\dfrac{(bh^3 - b_0 h_0^3)}{6h(bh - b_0 h_0)}$
	πab	$\dfrac{\pi ab^3}{4}$	$\dfrac{\pi ab^2}{4}$	$\dfrac{b}{4}$

Section	Area	I	W	i
	$\pi(ab - a_0 b_0)$	$\dfrac{\pi(ab^3 - a_0 b_0^3)}{4}$	$\dfrac{\pi(ab^3 - a_0 b_0^3)}{4b}$	$\dfrac{ab^3 - a_0 b_0^3}{4b(ab - a_0 b_0)}$
	$\pi t(a+b)$	$\dfrac{\pi b^2 t(3a+b)}{4}$	$\dfrac{\pi bt(3a+b)}{4}$	$\dfrac{b(3a+b)}{4(a+b)}$
	$0.433a^2$	$0.018a^4$	$\max\ \dfrac{a^3}{16}$ $\min\ \dfrac{a^3}{32}$	$\dfrac{a}{(8\sqrt{3})}$
	$\dfrac{bh}{2}$	$\dfrac{bh^3}{36}$	$\dfrac{bh^2}{24}$	$\dfrac{h}{12}$
	$\dfrac{(2b+b_1)h}{2}$	$\dfrac{6b^2 + 6bb_1 + b_1^2}{36(2b+b_1)}h^3$	$\dfrac{6b^2 + 6bb_1 + b_1^2}{12(3b+2b_1)}h^2$	$\dfrac{(6b^2 + 6bb_1 + b_1^2)h}{6(3b+2b_1)(2b+b_1)}$
	$BH + bh$	$\dfrac{BH^3 + bh^3}{12}$	$\dfrac{BH^3 + bh^3}{6H}$	$\dfrac{(BH^3 + bh^3)}{6H(BH + bh)}$

TABLE 5-1. (*Continued*)

	Section Area A	Moment of Inertia About x-x Axis I_x	Section Modulus About x-x Axis Z_x	$\dfrac{\text{Section Modulus}}{\text{Section Area}}$ $\propto \dfrac{\text{Strength}}{\text{Weight}}$
	$BH - bh$	$\dfrac{BH^3 - bh^3}{12}$	$\dfrac{BH^3 - bh^3}{6H}$	$\dfrac{(BH^3 - bh^3)}{6H(BH - bh)}$
	$aH + bd$	$I = \tfrac{1}{3}\left(Bc_1^3 - bh^3 + ac_2^3\right)$ $c_1 = \dfrac{1}{2}\dfrac{aH^2 + bd^2}{aH + bd}$ $c_2 = H - c_1$	$\dfrac{I}{c_2} = \dfrac{1}{3}\dfrac{\left(Bc_1^3 - bh^3 + ac_2^3\right)}{(H - c_1)}$ $\dfrac{I}{c_1} = \dfrac{1}{3}\left(Bc_1^3 - bh^3 + ac_2^3\right)$ $\dfrac{2(aH + bd)}{(aH^2 + bd^2)}$	$\dfrac{I/c_2}{(aH + bd)}$ or $\dfrac{I/c_1}{(aH + bd)}$

TABLE 5-2. Strength to Weight Ratios for Typical Sections

Shape	Area (in²)	Section Modulus / Area (strength-to-weight)
	5.00	1.43
	5.00	1.43
	3.75	0.78 (compression) 1.44 (tension)
	8.00	1.28
	7.07	1.03
	15.00	0.83
	5.00	0.83 (compression) 0.42 (tension)
	6.00	0.70

TABLE 5-2 (*Continued*)

Shape	Area (in²)	Section Modulus / Area (strength-to-weight)
5 inch (circular section)	19.63	0.63

thus producing a composite section. This increases bending strength considerably. Such composite sections can be of a variety of types, the simplest of which is shown in Figure 5-3.

If there is no slipping between the different sections of the composite at their common surface, then there will be continuity of strain as shown by the preceding strain diagram, and the whole composite beam will bend as one unit with a common radius of curvature. Typically, such composite sections are made of materials with widely differing moduli of elasticity (such as timber and steel). For the same strain, the stress carried by the stiffer element (steel) will be much higher than the stress in the other element (wood). Thus there will be a discontinuity in the stress diagram as shown in Figure 5-3d. Due to the discontinuity in the stress distribution pattern, the maximum bending moment that can be carried by the composite section is considerably increased.

Since the load carried by the steel section is higher than that for an equivalent timber section, for analytical purposes the cross section can be considered to consist entirely of wood with the width of the steel section increased by a factor of (E_s/E_w), as shown by Figures 5-3a and b. The width of the equivalent timber section is

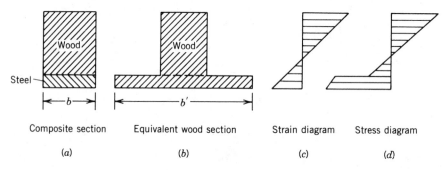

Composite section   Equivalent wood section   Strain diagram   Stress diagram

(*a*)   (*b*)   (*c*)   (*d*)

FIGURE 5-3. (*a*) Composite section of steel and wood, (*b*) equivalent all-wood section, (*c*) strain diagram for section, (*d*) stress diagram.

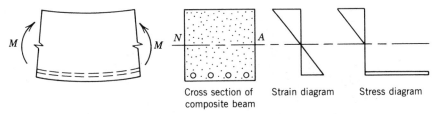

Cross section of   Strain diagram   Stress diagram
composite beam

FIGURE 5-4. Reinforced concrete section with strain and stress diagrams.

given by

$$b' = \left(\frac{E_s}{E_w}\right)b$$

where

E_s = modulus of elasticity of steel

E_w = modulus of elasticity of wood

b = width of the wood section

Because the moment of inertia and the section modulus of the composite section is much higher than that of an equivalent timber section, the composite section has greater bending strength.

Wood is a material that is stronger in compression than in tension, whereas steel has a comparatively high-tensile strength. The composite section can be configured so that the bending moment sets up compressive stresses in the wood and tensile stresses in the steel section such that both parts of the composite section are loaded to the optimum level.

Reinforced concrete is one of the common and outstanding examples of a composite section. Concrete is a material that has a high-compressive strength but is weak in tension and may in fact develop minute cracks that prevent it from carrying any tensile stresses. Generally, a portion of a concrete structure is reinforced with steel; such reinforcement is usually placed away from the neutral axis so that it is utilized to the best advantage. Figure 5-4 shows a reinforced concrete section under bending and the resulting strain and stress patterns.

5-5 SANDWICH CONSTRUCTION

Figure 5-5 shows a structural sandwich that is a type of laminar composite consisting of a lightweight core that is fixed between two thin, high-density face sheets.

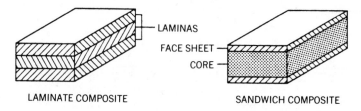

FIGURE 5-5. Laminar composites. [From Ref. 28, p. 17.]

Sandwich constructions are used in applications requiring rigidity and high strength-to-weight ratios. The load-carrying characteristics of a sandwich composite is similar to that of an *I*-beam that has a high-section modulus. Like the flanges of an *I*-beam, the face sheets of the sandwich composite carry the primary tensile and compressive loads. The core carries the primary shear load and maintains continuous support for the face sheets. The core thereby prevents buckling and wrinkling of the face sheets.

In practical applications, the shear rigidity of the core material must be sufficient to prevent deformations due to shearing of the structural sandwich. Because of lightweight and high strength, sandwich composites can be used to construct very rigid and lightweight structures. Typical sandwich constructions are shown in Figure 5-6. An exploded view of a honeycomb construction is depicted in Figure 5-7*a*.

For design analysis, the section properties of sandwich-type bearings and plates can be determined by making the following assumptions:

1. The core provides no stiffness to the structure.
2. The face sheet thickness is small.

For the sandwich composite shown in Figure 5-7*b*, the moment of inertia can be calculated by using the parallel axis theorem ($I = I_1 + Ad^2$) as

$$I = 2\left(\frac{bt^3}{12}\right) + 2bt\left[\frac{(c+t)}{2}\right]^2$$

For $c/t < 5$, the flexural rigidity, D, of the sandwich can be approximated as:

$$D = \frac{[E(h^3 - c^3)]}{[12(1 - v^2)]}$$

For $c/t > 5$, the equation of flexural rigidity can be written as

$$D = \frac{[Et(h + c)^2]}{[8(1 - v^2)]}$$

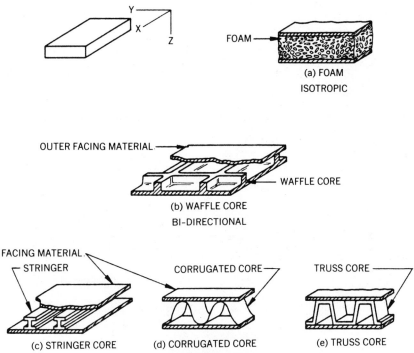

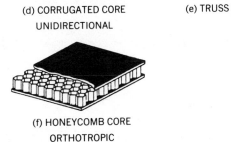

FIGURE 5-6. Typical sandwich constructions. [From Ref. 28, p. 18.]

Figure 5-8 depicts a simply supported sandwich panel subjected to a transverse load at its center. Due to the nature of the sandwich construction, the face sheets carry the major portion of the bending force, whereas the core carries the majority of the transverse-shear force induced by the load. The maximum bending stress of the system can be calculated as

$$\sigma_{max} = \frac{Mc}{I} = \frac{[(P/2)(L/4)]}{[btc(t + c/2)/(h/2)]}$$

or

$$\sigma_{max} = \frac{PL}{8btc}$$

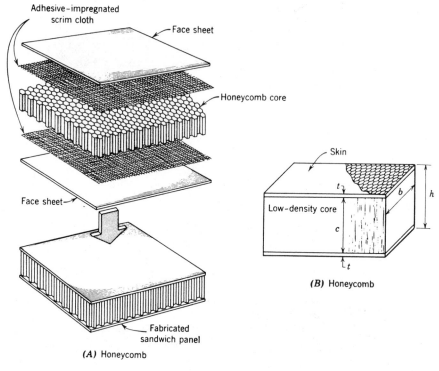

(A) Honeycomb

(B) Honeycomb

FIGURE 5-7. Honeycomb construction. [From Ref. 27, p. 296.]

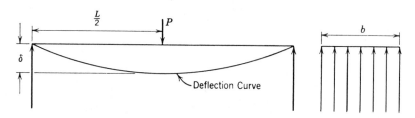

FIGURE 5-8. Panel subjected to center loading. [From Ref. 27, p. 301.]

The minimum stress is

$$\sigma_{\min} = \frac{PL}{8bth}$$

The maximum shear stress can be approximated over the average cross-sectional area of the sandwich and the core material as

$$\tau_{max} = \frac{S}{[b(h + c)/2]} = \frac{2S}{b(h + c)}$$

where S is the vertical shear load.

The maximum deflection, δ, of the panel occurs at its center and can be calculated, using Castigliano's theorem [27, p. 301].

5-6 ECCENTRIC AXIAL LOADS

An *eccentric axial load* is an axial load that does not pass through the centroidal axis of a cross section of the member. Figure 5-9 shows an eccentric axial load applied at a distance e from the centroidal axis. The eccentric load can induce bending stresses within the member. The induced bending moment on the member is *symmetrical* if the eccentric axial load can be resolved into an equivalent axial load through the centroidal axis and a bending moment about the neutral axis of the member. Figure 5-10 shows symmetric bending due to an eccentric axial load.

The bending moment on the member is *asymmetrical* if bending occurs about a neutral axis that does not pass through the centroid of the cross section. Figure 5-11 shows asymmetric bending due to an eccentric axial load. Figure 5-12 illustrates the case of noneccentric, symmetrical and asymmetrical loading.

Eccentric Axial Load with Symmetrical Bending

Consider the short block shown in Figure 5-13a. For the given eccentric loading, using superposition, we can construct the normal stress distribution in the member

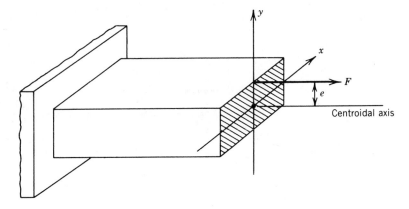

FIGURE 5-9. Eccentric axial loading.

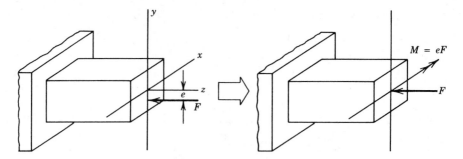

FIGURE 5-10. Symmetric bending.

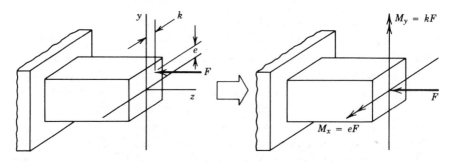

FIGURE 5-11. Asymmetric bending.

at section *b-b*. As a first step, we construct a free-body diagram of a portion of the block to the right of *b-b* as shown in Figure 5-13*b*. Force equivalence provides that the eccentric load P can be resolved into a compressive force $F = 100$ kN acting along the centroidal axis and a bending moment $M_u = Pe = 100$ kN (0.20 m) = 20 kNm about the neutral axis u of the member. Bending is therefore symmetric.

As the second step, we compute the principal moment of inertia with respect to the u axis:

$$I_u = \frac{bh^3}{12} = \frac{(0.25 \text{ m})(0.5 \text{ m})^3}{12} = 2.604 \times 10^{-3} \text{ m}^4$$

As a third step, we calculate the flexural stress σ_x due to the bending moment M_u:

$$\sigma_x = \frac{M_u v}{I_u} = 7.68v \frac{\text{MN}}{\text{m}^2}$$

where v represents the vertical distance from the neutral axis. The top section of

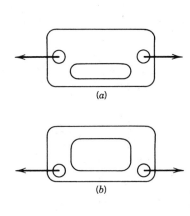

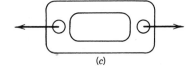

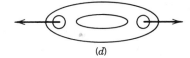

FIGURE 5-12. Effect of shape on bending. (*a*) non-eccentric, asymmetric, (*b*) eccentric, asymmetric, (*c*) and (*d*) noneccentric, symmetric.

the beam is subjected to tension and the section of the beam below the neutral axis is subjected to compression. The stress distribution due to bending is shown in Figure 5-14*a*.

As a final step, we calculate the normal stress σ_x due to the axial load *F*:

$$\sigma_x = \frac{F}{A} = \frac{100 \text{ kN}}{(0.25 \text{ m})(0.5 \text{ m})} = 0.80 \frac{\text{MN}}{\text{m}^2}$$

This compressive stress is constant across the section of the member and is pictured in Figure 5-14*b*.

The resultant stress distribution due to the eccentric load is the combined stress distribution due to the axial load *F* and the moment M_u. The total stress distribution across section *b-b* is shown in Figure 5-14*c*. The fiber at distance *d* is unstressed.

As another example of symmetric bending we consider the chain link in Figure 5-15. For the given eccentric loading, we need to calculate the normal stress distribution in the member at section *a-a*. First, we construct the free-body diagram

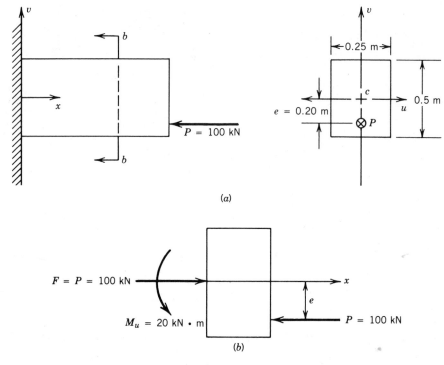

FIGURE 5-13. Eccentrically loaded block.

of a portion of the link to the left of section a-a as shown in Figure 5-15b. Force equivalence provides that the eccentric load T can be resolved into a tensile force $F = 15$ lbf and a bending moment

$$M_u = Pe = (15 \text{ lbf})(0.45 \text{ in}) = 6.75 \text{ lbf-in}$$

Bending is symmetric about the principal axis, u. The principal moment of inertia about the u axis is

$$I_u = \frac{bh^3}{12} = \frac{(0.25 \text{ in})(1.5 \text{ in})^3}{12} = 7.03 \times 10^{-2} \text{ in}^4$$

The flexural stress σ_x due to M_u is

$$\sigma_x = \frac{M_u v}{I_u} = 96v \text{ lbf/in}^2$$

The stress distribution due to the moment M_u is shown in Figure 5-16a.

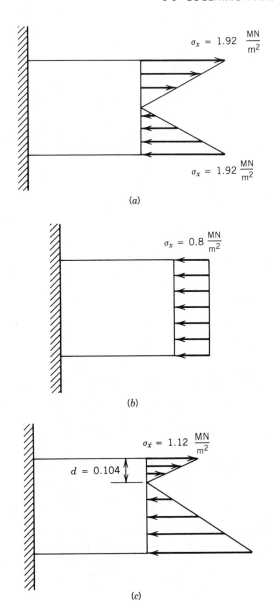

$\sigma_x = 1.92 \dfrac{\text{MN}}{\text{m}^2}$

$\sigma_x = 1.92 \dfrac{\text{MN}}{\text{m}^2}$

(a)

$\sigma_x = 0.8 \dfrac{\text{MN}}{\text{m}^2}$

(b)

$\sigma_x = 1.12 \dfrac{\text{MN}}{\text{m}^2}$

$d = 0.104$

(c)

FIGURE 5-14. Stress distribution across block.

The constant normal stress σ_x due to the tensile load F is given by

$$\sigma_x = \frac{F}{A} = \frac{15}{(0.25)(1.5)} = 40 \ \text{lbf/in}^2$$

The stress distribution due to F is shown in Figure 5-16b.

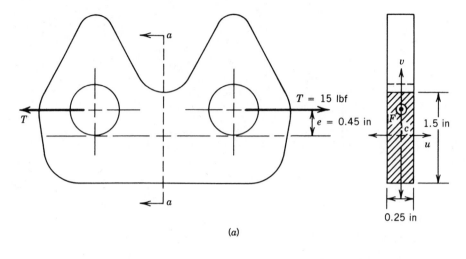

(a)

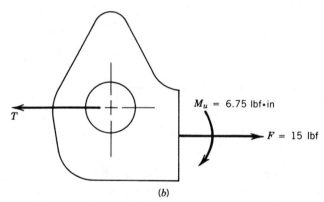

(b)

FIGURE 5-15. Silent chain link.

The combined stress distribution within the member at section *a-a* is illustrated in Figure 5-16*c*. The fiber of zero stress is located 0.333 in from the underside of the link.

Eccentric Axial Load with Asymmetric Bending [29]

Figure 5-17*a* shows a link with one slot. The link is loaded through end pins and carries a tensile force of 8000 lbf. Because of the slot, the neutral axis does not pass through the centroidal axis of section A-A, and consequently, section A-A is subject to eccentric loading as shown in Figure 5-17*b*. The eccentric load can be reduced to an axial load (*F*) of 8000 lbf through the neutral axis coupled with a bending moment (*M*) about the neutral axis of the cross section. The magnitude

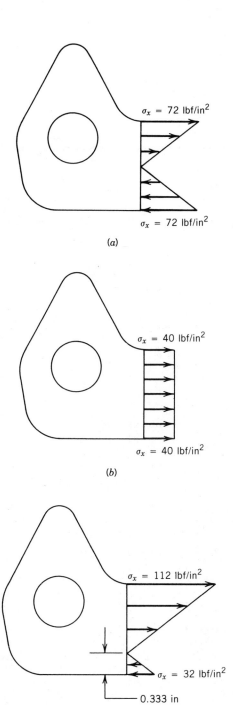

$\sigma_x = 72$ lbf/in^2

$\sigma_x = 72$ lbf/in^2

(a)

$\sigma_x = 40$ lbf/in^2

$\sigma_x = 40$ lbf/in^2

(b)

$\sigma_x = 112$ lbf/in^2

$\sigma_x = 32$ lbf/in^2

0.333 in

(c)

FIGURE 5-16. Stress distribution across chain link.

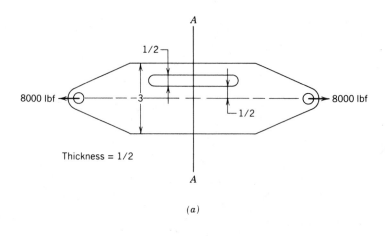

(a)

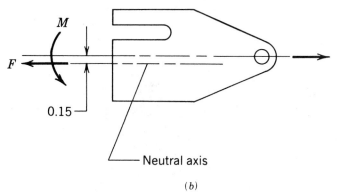

(b)

SECTION A-A

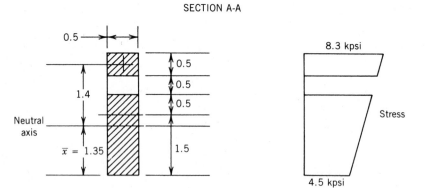

(c)

FIGURE 5-17. Asymmetric bending of a link. [From Ref. 29, p. 131.]

of the bending moment is equal to the product of the 8000 lbf tensile load and the eccentricity. To determine the eccentricity, we must find the position of the neutral axis. Summing moments about the base of the cross section in Figure 5-17c, we have

$$\bar{x} = \frac{[(2)(\frac{1}{2})(1)] + [(\frac{1}{2})(\frac{1}{2})(2.75)]}{[(2)(\frac{1}{2})] + [(\frac{1}{2})(\frac{1}{2})]} = 1.35 \text{ in}$$

The neutral axis is $(1.5 - 1.35) = 0.15$ in from the centerline.
    The bending moment is

$$M = (8000)(0.15) = 1200 \text{ lbf-in}$$

The total area moment of inertia I for the section is

$$I = [\tfrac{1}{2}(2)^3/12] + (\tfrac{1}{2})(2)(0.35)^2 + [(\tfrac{1}{2})(\tfrac{1}{2})^3/12] + (\tfrac{1}{2})(\tfrac{1}{2})(1.4)^2 = 0.951 \text{ in}^4$$

The stress due to bending is

$$\sigma_1 = \frac{My}{I} = \frac{1200y}{0.951} = 1261.83y$$

where y is the distance from the neutral axis.
    The uniform tensile stress due to the 8000 lbf axial load is given by

$$\sigma_2 = \frac{F}{A} = \frac{8000}{1.25} = 6400 \text{ psi}$$

where A is the area of the section. The resultant stress σ across the section at y is given by the sum of the bending and axial stresses

$$\sigma = \sigma_1 + \sigma_2 = 6400 + 1261.83y$$

where y is positive above the neutral axis (denoting tensile-bending stress) and negative below the neutral axis (denoting compressive-bending stress). The combined stress distribution across the section of the link is shown in Figure 5-17c.

6

TORSION

Certain cross-sectional shapes are best for torsional loading. A good designer would like to know why one shape is stiffer and stronger than another.

6-1 THE MEMBRANE ANALOGY

The membrane analogy can be used to visualize, intuitively, the stiffness and strength of a uniform straight bar of any cross section, twisted by couples applied at the ends, by considering certain properties of an inflated homogeneous membrane that is supported at its edges. The membrane analogy is developed from the fact that the differential equation for the stress function of a cross section for a torqued bar has the same form as the differential equation for the shape of a stretched membrane, originally flat, which is held at the edges of the cross section and inflated from the bottom. Figure 6-1 shows a membrane for an arbitrary cross section.

The membrane is simply supported at its edges and is subjected to a uniform lateral pressure. The partial differential equation for the deflection (shape) of the membrane in the z direction as a function of x and y can be written as [25, p. 304]

$$\frac{\partial^2 z}{\partial x^2} + \frac{\partial^2 z}{\partial y^2} = \frac{-p}{S} \qquad (6\text{-}1)$$

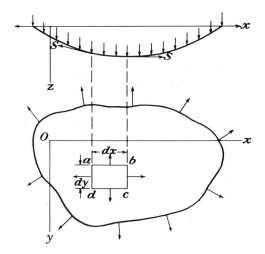

FIGURE 6-1. Homogeneous membrane supported at its edges, subjected to a uniform tension at the edges and a uniform lateral pressure. [From Ref. 11, p. 303.]

where p is the lateral pressure per unit area and S is the tensile force per unit length along the border(s) of the membrane. The differential equation of the shear stress function of a cross section for a torsionally loaded bar is given by [25, p. 295]

$$\frac{\partial^2 \phi}{\partial x^2} + \frac{\partial^2 \phi}{\partial y^2} = F = -2G\theta \tag{6-2}$$

where ϕ is called the stress function of x and y and θ is the angle of twist per unit length. Equations (6-1) and (6-2) are identical if

$$\frac{p}{S} = 2G\theta \tag{6-3}$$

If Eq. (6-3) is satisfied, the following statements are valid in accordance with the membrane analogy:

1. The slope of the membrane (h/t) is directly proportional to the shear stress in the direction perpendicular to the slope of the membrane deflection (see Figure 6-2).

2. The membrane contour lines are lines that follow the direction of the shear stresses.

3. The twisting moment is directly proportional to twice the volume under the membrane.

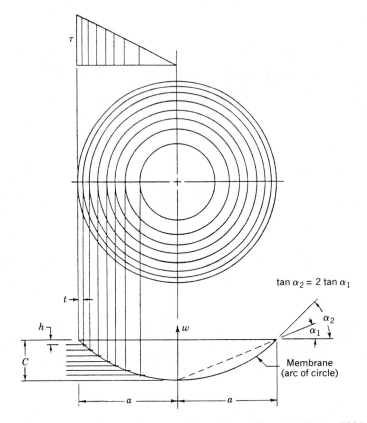

FIGURE 6-2. Membrane analogy of round bar. [From Ref. 31, p. 503.]

Calculation of Shear Stress Using the Membrane Analogy

The membrane analogy shows that the shear stress of an arbitrary cross section is proportional to the slope of the membrane's deflection curve. Mathematically

$$\tau = \frac{kh}{t}$$

where

h = incremental deflection of the membrane

t = incremental width

k = a constant of proportionality

The shear-stress equation for any cross section can be determined by making the following assumptions [11]:

1. The shear-stress distribution is linear across the shortest dimension of the cross section.

2. The profile of the membrane is parabolic in shape.

3. The deflection of the membrane is small.

If these assumptions are true for a given case, then the maximum shear stress can be written as

$$\tau_{max} = \frac{2M_t C}{VD}$$

where

M_t = twisting moment on the member

C = maximum deflection of the membrane

V = volume underneath the deflection curve

D = shortest dimension of the cross section

6-2 CIRCULAR VERSUS NONCIRCULAR CROSS-SECTIONAL SHAPES

It follows from the statements in Section 6-1 that if two cross sections have the same area, the section that is more nearly circular is the stiffer since that section creates the largest valume under the deflection curve, and hence would also produce the highest torque capacity.

To demonstrate this hypothesis, consider the circular shaft in Figure 6-3. Assuming that deflection of the membrane is small, the resultant pressure on the center portion (*mn*) of radius *r* is given by

$$P = \pi r^2 p \tag{6-4}$$

This total pressure, P, is balanced by the tensile force (S) per unit length which acts tangent to the membrane and in a direction defined by the slope of the membrane (dw/dr):

$$P = -2\pi r S \left(\frac{dw}{dr}\right) \tag{6-5}$$

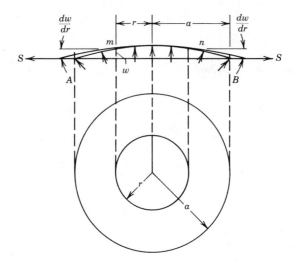

FIGURE 6-3. Circular shaft/membrane analogy. [From Ref. 38, p. 268.]

Equating Eqs. (6-4) and (6-5) gives

$$\frac{pr}{2S} = -\left(\frac{dw}{dr}\right) \tag{6-6}$$

The expression for P/S given in Eq. (6-1) can be substituted into Eq. (6-6) to obtain

$$-\left(\frac{dw}{dr}\right) = G\theta r = \tau \tag{6-7}$$

Equation (6-7) shows that the slope of the membrane is proportional to the shear stress.

   The twisting moment on the circular shaft can be found by first calculating the shaft deflection. Integration of Eq. (6-7) yields

$$w = \int_r^a G r\theta \, dr = \frac{G\theta(a^2 - r^2)}{2} \tag{6-8}$$

The volume under the deformed membrane is then found by

$$V = \int_0^a 2\pi r \, drw = G\theta\left(\frac{\pi a^4}{4}\right) = \frac{G\theta J}{2} \tag{6-9}$$

where $J = \pi a^4/2$ is the area polar moment of inertia. The twisting moment is proportional to twice the volume under the membrane. The expression for the twisting moment of a solid circular shaft is therefore

$$T = G\theta J = G\theta \left(\frac{\pi a^4}{4} \right) \tag{6-10}$$

An expedient method to determine the maximum shear stress of the solid shaft is to approximate the deflection volume as

$$V = \tfrac{1}{2}\pi a^2 C$$

where a is the radius of the circular section. The maximum shear stress for the circular section is given by

$$\tau_{\max} = \frac{2T}{\pi a^3} \tag{6-11}$$

As an example of the torque capacity of a noncircular cross section, consider the thin plate shown in Figure 6-4. The volume underneath the parabolically shaped membrane can be approximated by

$$V = \frac{2BDC}{3} \tag{6-12}$$

where B is the length and D is the width of the cross section. The shear stress is given by

$$\tau_{\max} = \frac{3T}{BD^2} \tag{6-13}$$

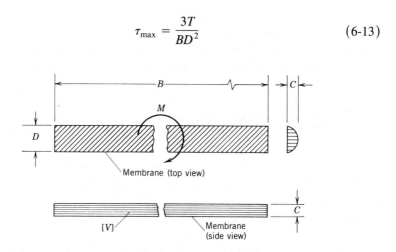

FIGURE 6-4. Thin plate. [From Ref. 31, p. 504.]

A comparison between the maximum shear stresses of the solid circular and the rectangular cross sections shows that the torque capacity of the circular cross section is significantly greater than that of the noncircular section. For example, if the areas of both sections are assumed to be equal to 1.0 in^2, then for the circular section

$$a = 0.564 \text{ in}$$

and

$$\tau_{max} = \frac{2T}{\pi a^3} = 3.55T \qquad (6\text{-}14)$$

For a narrow rectangular section

$$B = 6 \text{ in}$$

where

$$D = \frac{1}{6} \text{ in}$$

and

$$\tau_{max} = \frac{3T}{BD^2} = 18T \qquad (6\text{-}15)$$

Inspection of Eqs. (6-14) and (6-15) shows that the rectangular section carries a shear stress that is approximately five times greater than that for the circular section.

6-3 EFFECT OF INWARD AND OUTWARD PROTRUDING CORNERS

Stresses at or near outward-protruding corners in the cross-sectional shape are low; stresses at or near inward-protruding corners are high. This statement follows from the fact that the slope of the membrane is smallest at outward-protruding corners and largest at inward-protruding corners. Consider the stress concentration of a notched circular shaft in torsion. Figure 6-5 shows that reentrant sharp corners cause high-stress concentration effects. Figure 6-5 shows that projecting corners have no critical stress effects. The slope of the membrane is large at inward-

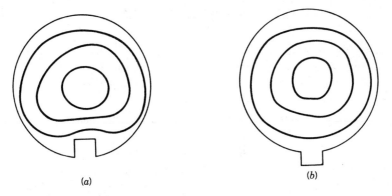

(a)   (b)

FIGURE 6-5. Notch effects in torsion.

protruding corners since the membrane must rise faster to get to the highest point in the membrane.

It also follows that substantial stress concentrations occur at the inner corners of angle sections, channels, I-sections, and at the fillets of the inside corners of keyways. But a fillet will clearly have a lower stress than a square corner. And, in general, the greatest stresses in a given section occur at an outer boundary, a boundary adjacent to the thickest portion or the greatest inscribed circle, and in particular, at the boundary that is most concave algebraically when viewed from outside the section. The minimum stress would be zero at outward-protruding corners or near centers of inscribed circles where the membrane slope is zero.

Figure 6-6 pictures the cross section of a straight bar. The largest stress would be found at the boundary point A where the slope of the membrane would be the greatest.

In regard to stiffness, a small, narrow inward or outward protrusion has little effect since small boundary protrusions have little effect on volume, and the volume under the membrane is directly proportional to the stiffness of the section.

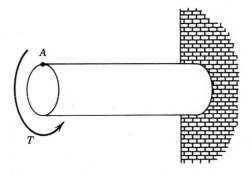

FIGURE 6-6. Cross section of extruded bar.

6-4 CLOSED VERSUS OPEN CROSS SECTIONS

Consider now those cross-sectional shapes shown in Figure 6-7. The membrane analogy states that the torsional capacity of a cross section is directly proportional to the volume under a bubble membrane created by using the outer perimeter of the cross section to hold the outer edge of the bubble. The figure shows six different cross sections all with the same cross-sectional area. The volume produced under a bubble tied down at the perimeter of each cross section with a pressure underneath would be approximately the same, thus the membrane analogy points out that the torque capacity for each of the six cross-sectional shapes would almost be identical.

Referring to Figure 6-8, we find that the membrane analogy in this case would reveal that the resistance of an I-beam to torsion would be equal to the torsional resistance of the center web plus two times the torsional resistance of one flange. In other words, R, the torsional resistance of the I-beam, would be equal to R_3, the resistance of the center web, plus two times R_2, the torsional resistance of one flange.

Figure 6-9 gives the torsional resistance for a plate that is 2 in long and 0.055 in thick and the torsional resistance for an I-beam that is 2 in wide and 2 in deep. If the torsional load on the I-beam is the same as that on the plate, the figure shows that the I-beam is about three times stiffer than the plate. Figure 6-10 shows five cross-sectional shapes: a plate, a channel, a slotted circular tube, a hollow tube, and a square hollow tube. The closed shapes (d and e) are much stronger and

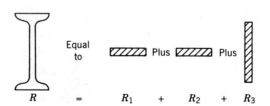

FIGURE 6-7. Poor sections for torsional loading—open sections. [From Ref. 2, p. 3.6-4.]

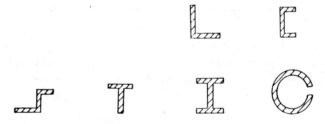

FIGURE 6-8. Torsional resistance of an I-beam. [From Ref. 2, p. 3.6-3.]

Angle of Twist		
All loadings identical	$t = .055$ 	$t = .055$ 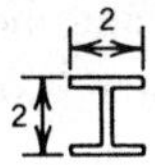
Conventional method J Polar moment of inertia	.065°	.007°
Method using R Torsional resistance	21.8°	7.3°
Actual twist	22°	9.5°

FIGURE 6-9. Results of twisting an I-beam made of three identical plates. [From Ref. 2, p. 3.6-3.]

Angle of Twist					
	a	b	c	d	e
All loadings identical	$t = .060$ 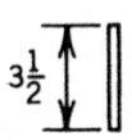	$t = .060$ 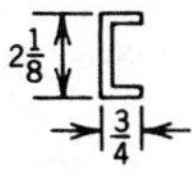	$t = .060$ 	$t = .060$	$t = .060$ 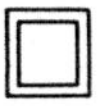
Conventional method J Polar moment of inertia	.01°	.006°	.04°	.04°	.045°
Method using R Torsional resistance	9.5°	9.7°	10°	.04°	.06°
Actual twist	9°	9.5°	11°	Too small to measure	Too small to measure

FIGURE 6-10. Several open and closed sections. [From Ref. 2, p. 3.6-3.]

stiffer in torsion than the open sections (a), (b), and (c), as indicated by the values of the torsional resistance R. The volume created underneath the membrane for the I-beam, the channel, and the slit tube would be approximately the same. In this case, the torsion resistance of each shape would be identical. However, the resistance of the closed shapes, the hollow tube and the square tube, are extremely large as compared to the open shapes. The actual twists for the closed shapes are indeed too small to measure.

The membrane analogy can be applied to predict the relative torsional stiffness of a solid versus a hollow shaft. The actual volume under the membrane for a hollow-shaft cross section is equal to the volume found under a solid shaft minus the volume found under a solid shaft the size of the center hole. This difference in volume is much larger than expected, and hence the torque capacities of the closed cross sections are extremely high. The maximum shear-stress equation for a hollow shaft of outer diameter D and inner diameter d can be written as [31, p. 508]

$$\tau_{max} = \frac{16TD}{\pi(D^4 - d^4)}$$

For a cross-sectional area of 1 in^2 and an outer diameter (D) of 5 in, $d = 4.943$ in and the maximum shear stress is

$$\tau_{max} = 0.913T$$

The maximum shear stress of the hollow shaft is smaller than that of the solid shaft $(\tau_{max} = 3.55T)$ for equal circular cross-sectional areas. This shows that for shafts

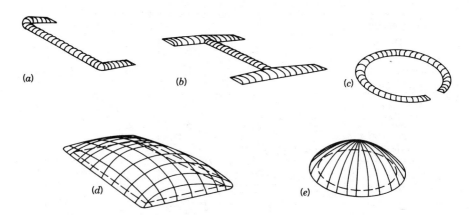

FIGURE 6-11. Membranes for five cross sections. [From Ref. 2, p. 3.6-16.]

having the same shear strength the torque capacity of the hollow shaft is approximately 3.9 times greater than the capacity of the solid shaft. Consequently, closed sections should be employed when a high-torsional resistance is desired.

Figure 6-11 pictures membranes for five common structural sections. Comparison of (d) and (e) with (a), (b), and (c) readily indicates that closed shapes are many times more resistant to twist than open shapes are.

7

CONTACT STRESS

A state of triaxial stress often arises when two bodies having curved surfaces are pressed against one another. Area contact results between the two surfaces and triaxial stresses are developed in the bodies. These stresses are defined as contact stresses. Often in engineering practice, the contact-stress pattern between two solids must be uniform in order to achieve the most efficient and durable design.

7-1 HERTZ EQUATIONS FOR THE GENERAL CASE OF CONTACT

The Hertz theory of elastic contact between two bodies is often utilized by the designer for predicting the area of contact and the stresses between the surfaces of the bodies. The Hertz theory is based on the following assumptions [32, p. 26]:

1. The surfaces of the bodies are frictionless.
2. The bodies are isotropic and homogeneous.
3. The surfaces are topographically smooth and continuous.
4. The surface tractions are due to contact forces only. Adhesion is negligible.
5. The profiles of the bodies can be represented by a second-degree surface.
6. The stresses and displacements may be deduced from the small strain theory of elasticity applied to a linearly elastic half-space.

The general case of the compression of two elastic bodies in contact is shown in Figure 7-1. Following the development given in Ref. 11, the tangent plane at the point O is defined as the xy plane. Locations M and N are designated points that are a distance r from the normal through the tangent plane. The initial separation of the points M and N from the tangent plane are denoted by the axes z_1 and z_2. The surfaces of the bodies near the point of contact are described by

$$z_1 = A_1 x^2 + A_2 xy + A_3 y^3 \tag{7-1}$$

$$z_2 = B_1 x^2 + B_2 xy + B_3 y^2 \tag{7-2}$$

where A and B are constants and provided that the following assumptions are valid [11]:

1. The small quantities of higher order in Eqs. (7-1) and (7-2) are neglected.
2. The surface adjacent to the point of contact is semicircular and may be modeled as a second-degree surface.

The distance between points M and N is therefore

$$z_1 + z_2 = (A_1 + B_2)\, x^2 + (A_2 + B_2)\, xy + (A_3 + B_3)y^2 \tag{7-3}$$

By taking a trajectory where xy would equal zero, we find that Eq. (7-3) becomes

$$z_1 + z_2 = Ax^2 + By^2$$

where the constants A and B are determined by the radii of curvature of the contact surfaces and by the angle between the planes of principal curvature of the two surfaces.

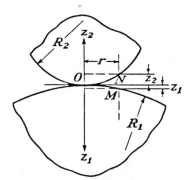

FIGURE 7-1. Two elastic bodies in contact. [From Ref. 11, p. 410.]

When a load is applied, the displacement of point M in the z_1 direction can be denoted as w_1. Likewise, the displacement of point N can be defined as w_2.

If the tangent plane xy is fixed and bodies 1 and 2 are subjected to local compression, then points M and N will approach one another by some amount α. Then, by geometry,

$$w_1 + w_2 = \alpha - (z_1 + z_2)$$

or

$$w_1 + w_2 = \alpha - Ax^2 - By^2 \qquad (7\text{-}4)$$

Now consider the local deformation at the surface of contact. The sum of the displacements w_1 and w_2 for points on the contact surface is given by [11]

$$w_1 + w_2 = \left[\frac{(1 - \nu^2)}{\pi E_1} + \frac{(1 - \nu^2)}{\pi E_2} \right] \iint \frac{q\, dA}{r} \qquad (7\text{-}5)$$

where $q\, dA$ is the pressure acting on an infinitesmally small element on the surface of contact and r is the distance of this element from the point referenced.

Equating Eqs. (7-4) and (7-5) yields [11, p. 415]

$$(k_1 + k_2) \iint \left(\frac{q\, dA}{r} \right) = \alpha - Ax^2 - By^2 \qquad (7\text{-}6)$$

where k_1 and k_2 denote the constant coefficients in Eq. (7-4).

Hertz determined that the distribution of pressures necessary to satisfy Eq. (7-6) can be determined by assuming that the intensity of pressure q over the surface of contact is represented by the semiaxes a and b of a semiellipsoid constructed over the surface of contact [11]. Therefore the maximum pressure (q_0) occurs at the center of the contact surface. The magnitude of the average pressure at the surface of contact is given by

$$P = \iint q\, dA = \frac{2\pi abq_0}{3}$$

and the maximum pressure is thus

$$q_0 = \frac{3P}{2\pi ab}$$

The maximum pressure on the surface of contact is 50 percent higher than the

TABLE 7-1. Coefficients of Semiaxes [11, p. 416]

$\theta =$	30°	35°	40°	45°	50°	55°	60°	65°	70°	75°	80°	85°	90°
$m =$	2.731	2.397	2.136	1.926	1.754	1.611	1.486	1.378	1.284	1.202	1.128	1.061	1.000
$n =$	0.493	0.530	0.567	0.604	0.641	0.678	0.717	0.759	0.802	0.846	0.893	0.944	1.000

average pressure. For calculations, the semiaxes a and b of the ellipsoid are given by [11]

$$a = m^3 \left[\frac{3\pi P(k_1 + k_2)}{4(A + B)} \right]$$

$$b = n^3 \left[\frac{3\pi P(k_1 + k_2)}{4(A + B)} \right]$$

The coefficients m and n are functions of A and B and can be determined from Table 7-1. The relationship between θ and A and B is given by

$$\cos \theta = \frac{(B - A)}{(A + B)}$$

By assuming that the x and y axes of the tangent plane are in the same direction as the semiaxes a and b, we determine the principal stresses at the center of contact as [11, p. 417]

$$\sigma_x = -2\nu q_0 - (1 - 2\nu)q_0 \left[\frac{b}{(a + b)} \right]$$

$$\sigma_y = -2\nu q_0 - (1 - 2\nu)q_0 \left[\frac{a}{(a + b)} \right]$$

$$\sigma_z = -q_0$$

where ν is Poisson's ratio.

7-2 CONTACT STRESSES BETWEEN PARALLEL CYLINDERS

Consider the pressure distribution across the contact area of the two parallel cylinders shown in Figure 7-2. The maximum pressure on the middle of the strip was determined by Hertz [41, p. 12]:

$$p_{max} = 0.418 \left[\frac{PE}{L} \left(\frac{1}{R_1} + \frac{1}{R_2} \right) \right]^{1/2} \tag{7-7}$$

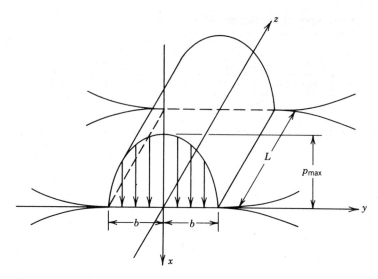

FIGURE 7-2. Pressure distribution across the strip of contact surface of two cylinders. [From Ref. 41, p. 12.]

where

R_1 and R_2 = radii of the two cylinders

L = contact length

E = modulus of elasticity

P = contact force

Equation (7-7) was developed with the following assumptions [11]:

1. The bodies in contact are isotropic.
2. The direction of the loading is perpendicular to the contact surface.
3. The proportional limit of the material is not exceeded.
4. The radii of curvature of the contact area (R_1 and R_2) are large as compared with the dimensions of the contact area.

In 1924, Thomas and Hoersch determined the principal stress distributions in the contact zone along the line of symmetry of the cylinders [41]. The distributions of the stress components σ_x, σ_y, σ_z, and τ_{max} are given in Figure 7-3. The stress τ_{max} is the shear stress on a 45° plane from the y axis. The shear stress as a function

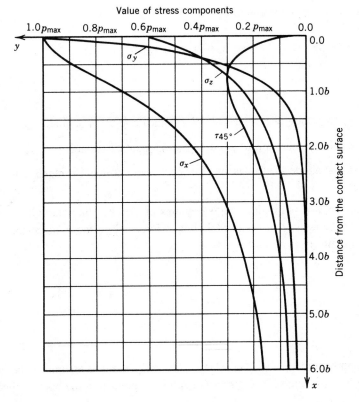

Value of stress components

FIGURE 7-3. Stress components in rolling cylinders. [From Ref. 41, p. 13.]

of σ_x and σ_y is [41, p. 13]

$$\tau_{max} = \frac{(\sigma_x - \sigma_z)}{2}$$

The magnitudes of σ_x, σ_y, and σ_z decrease as the distance from the contact surface increases. The maximum shear stress, however, occurs at a distance of $1.0b$ from the contact surface.

7-3 CONTACT-STRESS NUMERICAL METHOD

The Conry and Seireg [52] numerical method can be used to determine the contact-stress distribution for geometries where the Hertz theory would not be applicable. Conry and Seireg show that when two bodies are in contact, their mutual displacements at point i can be written as

$$w_i + z_i - \alpha \geqq 0 \qquad (7\text{-}8)$$

where

w_i = magnitude of displacement for point i due to an applied load

z_i = initial separation of point i from the contact surface

α = magnitude of displacement of point i due to local compression

If the contact area is divided into N' squares of side Δx, the total contact force P is equal to the sum of the elemental contact forces F_j on those elemental squares. The total contact force can be expressed as

$$P = \sum_{j=i}^{N} F_j \qquad (7\text{-}9)$$

where $j \leqq N \leqq N'$.

If the number of squares (N) is large, then the pressure is assumed to be constant within a given element. The displacement of the point i on the body is then

$$w_i = \sum_{j=i}^{N} a_{ij} F_j \qquad (7\text{-}10)$$

where a_{ij} is the influence coefficient derived from the theory of elasticity.

A formal statement of the contact problem can be made if a variable (Y_i) is introduced to make Eq. (7-8) an equality.

The variable Y_i is defined as the separation between the points N that do not contact. Equation (7-8) then becomes

$$-\sum_{j=i}^{N} a_{ij} F_j + \alpha + Y_i = z_i, \qquad i = 1, 2, \ldots, N$$

$$\sum_{j=i}^{N} F_j = P$$

$$\sum_{j=i}^{N} F_j Y_j = 0 \qquad (7\text{-}11)$$

where

$$F_i \geqq 0, \quad Y_i \geqq 0, \quad \alpha \geqq 0, \quad i = 1, 2, \ldots, N$$

Equations (7-11) can be reformulated as a mathematical programming problem that can be solved numerically [44].

Numerical methods for contact stress and pressure analysis can be employed for systems of any geometry. For example, consider the case of two identical steel spheres in contact (Fig. 7-4). The total deflection δ of the spheres under an applied load of 128.5 lbf is 4×10^{-4} in. The numerical results were obtained by dividing the expected contact area into nine strips, each strip having nine elements [40, p. 6]. The Chen [40] numerical method and the Hertz method were both applied with almost identical results. The approximate pressure distribution across the contact surface is shown in Figure 7-4.

As another example, consider the axial pressure distribution of a crowned roller pressed against a shaft. The system is shown in Figure 7-5. The parameters of the problem are given by [40, p. 6]

$$R_1 = 0.1966 \text{ in}$$

$$R_2 = 1.082 \text{ in}$$

$$L_s = 0.189 \text{ in}$$

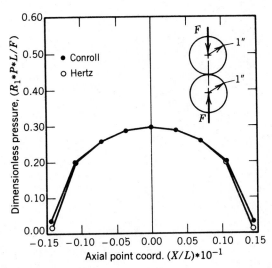

FIGURE 7-4. Pressure distribution for two identical spheres in contact. [From Ref. 40, p. 6.]

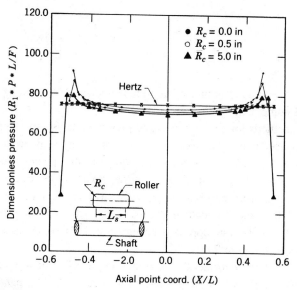

FIGURE 7-5. Axial pressure distribution, crowned roller loaded against a shaft, $F = 82$ lbf. [From Ref. 40, p. 6.]

$$R_c = 0, 0.5, \text{ and } 5 \text{ in}$$

$$E_1 = E_2 = 30 \times 10^6 \text{ psi}$$

$$\nu_1 = \nu_2 = 0.3$$

where

R_1 = roller radius

R_2 = shaft radius

L_s = straight length of roller

R_c = crowned radius at both edges of the roller

E = elastic modulus

ν = Poisson's ratio

The numerical results were obtained by dividing the expected contact area into 21 strips of 5 elements each. The axial pressure distribution between the roller and the shaft under a load of 82 lbf is shown in Figure 7-5. The pressure and stress is highest at the ends of the roller.

7-4 CONTACT STRESSES IN INDENTED STRIPS AND SLABS

The contact-stress distribution between strips or slabs can be determined by numerical methods [41, Hartnett]. In this section, we determine the contact-stress distributions for a plane strip compressed by a rigid, axially loaded, flat-ended indenter for two cases. In the first case, the effects of friction between the contact surfaces are ignored and the indented surface expands laterally without restraint. The second case considers the effects of friction, and the contact surfaces are coarse enough to restrain completely the lateral expansion of the indented surface.

The method of solution for determining the pressure distribution under the indenter can be described with reference to Figure 7-6. The strip of infinite length in the x direction is compressed from two sides by a uniformly distributed force, P, over the length 4ϵ. The elemental pressures $p_1, p_2, \ldots, p_n$ can be determined by the method of superposition. If contact takes places over the total length (4ϵ) of the indenter, then the vertical deflection of the slab is a function of the elemental pressures (p_i) over n lengths. The vertical deflection is constant along the total length in the case of a flat punch. The solutions of the n simultaneous linear equations yield the values of the elemental pressures $(p_1, p_2, \ldots, p_n)$ which approximate the actual pressure distribution under the indenter.

Figure 7-7 depicts the pressure distribution of the indented slab for the frictionless case. Figure 7-8 shows the pressure distribution of the slab for the case of total "lateral adhesion." As expected, the normal contact stress distributions for

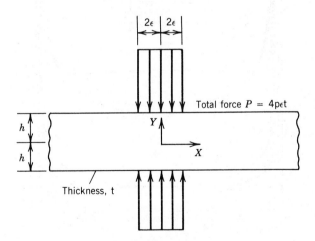

FIGURE 7-6. Elastic strip compressed by uniform pressure. [Reprinted with permission from *International Journal of Engineering Science*, Vol. 4, p. 345 H. Conway, S. Vogel, K. Farnham and S. So, "Normal and Shearing Contact Stresses in Indented Strips and Slabs." Copyright 1966, Pergamon Press, Ltd.]

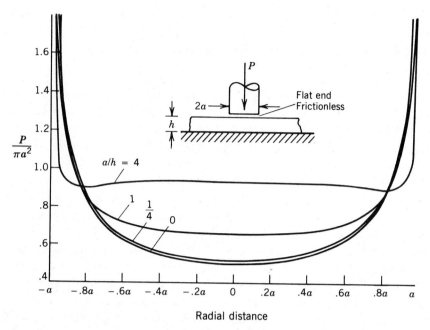

FIGURE 7-7. Normal stresses in indented slabs (frictionless case). [Reprinted with permission from *International Journal of Engineering Science*, Vol. 4, p. 350 H. Conway, S. Vogel, K. Farnham and S. So, "Normal and Shearing Contact Stresses in Indented Strips and Slabs." Copyright 1966, Pergamon Press, Ltd.]

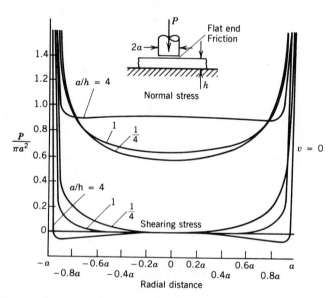

FIGURE 7-8. Normal and shearing contact stresses in indented strips and slabs. [Reprinted with permission from *International Journal of Engineering Science*, Vol. 4, p. 351 H. Conway, S. Vogel, K. Farnham and S. So, "Normal and Shearing Contact Stresses in Indented Strips and Slabs." Copyright 1966, Pergamon Press, Ltd.]

the frictionless and the adhesion cases are very similar because friction forces have minimal effect on the pressure in the normal direction.

The shear stress and normal stress of the coarse slab are greatest near the periphery of the indenter. The shear stress was zero at the geometric center of the coarse slab, due to the absence of lateral movement of material.

7-5 CONTACT STRESS OF TYPE CHARACTERS

The indentation of a flat type character into a platen is considered to be a special case of the indenter and slab contact problem.

The Conry and Seireg numerical method can be applied to determine the contact stress of type characters. The method can be applied under the following assumptions [44, p. 658]:

1. The platen is represented by a linearly elastic, isotropic, and homogeneous half-space.
2. The print character is much more rigid than the platen so that the deflection of the character is negligible.

The graphical results presented in Figure 7-9 show the pressure distribution on steel type characters (E_1 = 207 GPa, ν_1 = 0.3) pressed into a hard rubber platen (E_2 = 620 MPa, ν_2 = 0.5). Figure 7-9a shows the contact-pressure profile of a capital "I" and Figure 7-9b shows the pressure profile of a capital "T" character. In both profiles, the pressure increases toward the perimeter of the character. This phenomenon, as seen in Section 7-4, is common in flat punch problems.

The optimal design for a type character or a common punch is one that produces an efficient or uniform contact-stress pattern. Such a design gives longer punch life and, in the case of the type character, better print quality. Uniform-pressure distribution across the contact surface can be achieved through the optimization of the initial separation of the bodies (z_i). Generally, the optimal value of z_i for each element on the contact surface compensates for the high stresses at the perimeter of the punch by increasing the distance between the outer sections of the indenter and the platen, thereby decreasing the relative pressure. The distance (z_i) between the central sections of the punch (sections of lower stress) are modified to increase the pressure across the midsection of the contact areas. The optimal shapes for the capital type characters "I" and "T" and their improved stress distributions are shown in Figure 7-10. Note that the resultant shapes of the characters are no longer flat but are now double convex configurations.

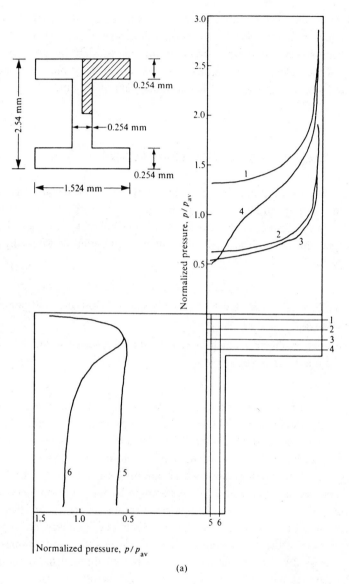

FIGURE 7-9. Contact stresses in type characters. [From Ref. 44, p. 660. Copyright 1978 by International Business Machines Corporation; reprinted by permission.]

7-6   GEAR TOOTH OPTIMIZATION TO MINIMIZE WEAR

Figure 7-11 shows a gear coupling. Gear couplings often employ the straight-sided tooth form shown in Figure 7-12. The tooth form is cut with a standard involute profile and a 14.5° pressure angle. The major drawback of the straight involute

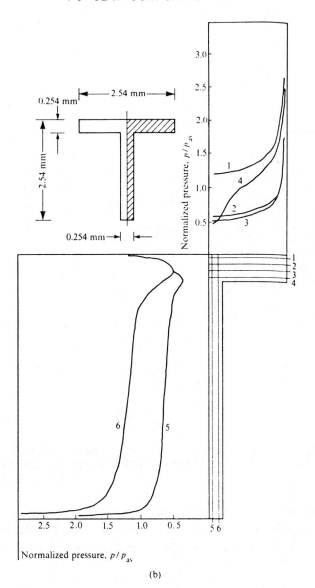

(b)

FIGURE 7-9. (Continued)

tooth form is the Hertzian contact between the flank of the tooth and the load-bearing surface.

Frequently, misalignment of the gear teeth occurs when the coupling is in operation and the point of contact shifts to the end of the straight tooth. Misalignment of a straight-sided gear tooth is shown in Figure 7-13a. Misalignment causes extremely high Hertz contact stresses at the ends of the teeth where a tooth is weakest.

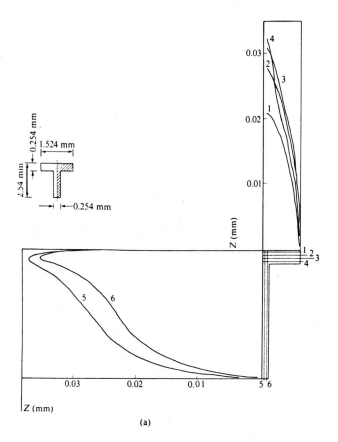

FIGURE 7-10. Optimized contact stresses in type characters. [From Ref. 44, p. 663. Copyright 1978 by International Business Machines Corporation; reprinted by permission.]

A higher contact stress accelerates the wear of the gear teeth and reduces the life of the coupling. The Hertz stress in gear teeth is based on the general Hertz equation for a cylinder on a plane. The equation for Hertz compressive stress is [35, p. 5]

$$\sigma = \left[\frac{PE}{D} \right]^{1/2}$$

where P is the load per unit length of the cylinder in contact, E is the modulus of elasticity and D is the diameter of the cylinder.

The state of the art in gear-coupling design suggests a fully crowned tooth form to reduce the coupling wear caused by excessively high Hertz contact stresses. Misalignment of a crowned tooth is shown in Figure 7-13*b*. A fully crowned tooth

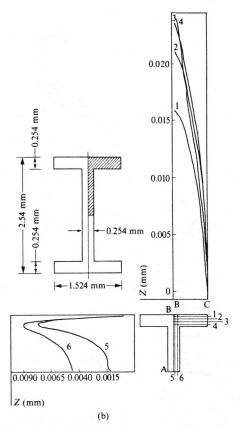

0.020

0.015

0.010

0.005

0

← 0.254 mm

2.54 mm

0.254 mm

→ 0.254 mm ←

|← 1.524 mm →|

Z (mm)

3 4
2

1

B B   C

0.0090 0.0065 0.0040 0.0015

6   5

B B   C
1 2
4   3

A
5 6

Z (mm)

(b)

FIGURE 7-10. (*Continued*)

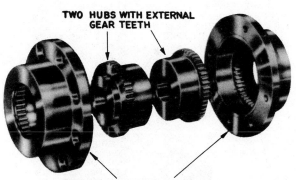

TWO HUBS WITH EXTERNAL
GEAR TEETH

TWO SLEEVES WITH INTERNAL
GEAR TEETH

BASIC COMPONENTS OF DOUBLE ENGAGEMENT
GEAR TYPE COUPLING

FIGURE 7-11. Gear coupling. [From Ref. 35, p. 4.]

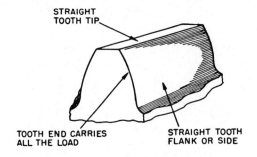

STRAIGHT
TOOTH TIP

TOOTH END CARRIES
ALL THE LOAD

STRAIGHT TOOTH
FLANK OR SIDE

FIGURE 7-12. The ordinary straight-sided external tooth form. [From Ref. 35, p. 5.]

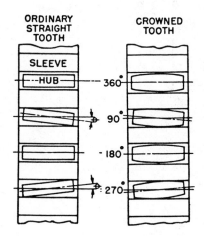

ORDINARY
STRAIGHT
TOOTH

CROWNED
TOOTH

SLEEVE

HUB

360°

90°

180°

270°

FIGURE 7-13. Gear tooth misalignment. [From Ref. 35, p. 5.]

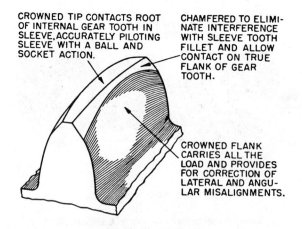

CROWNED TIP CONTACTS ROOT
OF INTERNAL GEAR TOOTH IN
SLEEVE, ACCURATELY PILOTING
SLEEVE WITH A BALL AND
SOCKET ACTION.

CHAMFERED TO ELIMI-
NATE INTERFERENCE
WITH SLEEVE TOOTH
FILLET AND ALLOW
CONTACT ON TRUE
FLANK OF GEAR
TOOTH.

CROWNED FLANK
CARRIES ALL THE
LOAD AND PROVIDES
FOR CORRECTION OF
LATERAL AND ANGU-
LAR MISALIGNMENTS.

FIGURE 7-14. The fully crowned tooth form. [From Ref. 35, p. 6.]

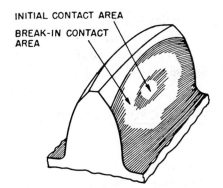

INITIAL CONTACT AREA

BREAK-IN CONTACT AREA

FIGURE 7-15. Fully crowned tooth wear pattern. [From Ref. 35, p. 7.]

form is shown in Figure 7-14. The crowned tooth has three basic parts: the crowned flank, the crowned tip, and the chamfered tip. The crowned flank eliminates the point loading on the tips of the tooth by increasing the area of contact. The increase from point contact to area contact thereby reduces the Hertz compressive stresses on the flank of the tooth.

The crowned tip contacts the root of a mating gear tooth, and this reduces the

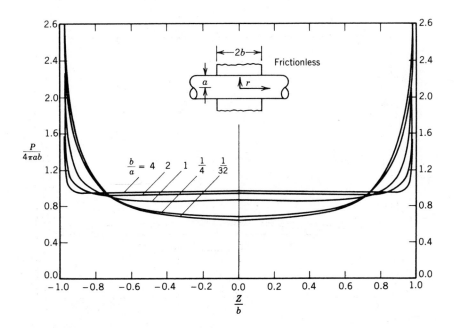

FIGURE 7-16. Pressure distributions between a circular cylinder and a rigid sleeve for various values of b/a (frictionless contact). [Reprinted with permission from *International Journal of Engineering Science*, Vol. 5, p. 542. H. Conway and K. Farnham, "Contact Stresses Between Cylindrical Shafts and Sleeves." Copyright 1967, Pergamon Press, Ltd.]

possibility of coupling misalignment [35]. The chamfered tip permits full contact on the flanks of the mating gear teeth. An improved misalignment pattern for a fully crowned tooth is shown in Figure 7-13. An improved wear pattern due to the increased contact area of a fully crowned tooth is shown in Figure 7-15.

7-7 CONTACT STRESS BETWEEN PRESS-FITTED CYLINDRICAL SHAFTS AND SLEEVES

Nonuniform contact-stress distributions arise when a rigid sleeve is press-fitted or shrunk onto a circular cylinder. In order to determine the nonuniformity of the stress distribution, we consider two cases. In the first case, the effects of friction are ignored, and in the second case, the effects of friction cause complete "adhesion" between the sleeve and the cylinder. In both cases, the following assumptions are made [42]:

1. The rigidity of the sleeves is much greater than that of the cylinder.
2. The cylinder is very long so that the end effects are negligible.

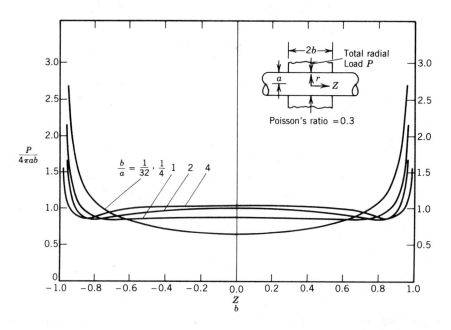

FIGURE 7-17. Normal and shearing contact stresses in press-fitted shafts and sleeves. [Reprinted with permission from *International Journal of Engineering Science*, Vol. 5, pp. 544, 545. H. Conway and K. Farnham, "Contact Stresses Between Cylindrical Shafts and Sleeves." Copyright 1967, Pergamon Press, Ltd.]

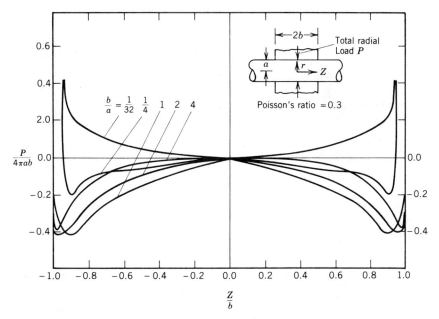

FIGURE 7-17. (*Continued*)

The numerical method of analysis for the stated cases involved the replacement of the normal and shearing contact-stress distributions by elemental pressures that are constant over large numbers of infinitesimally small lengths. The approximate stress distribution was then determined by applying the boundary conditions at discrete points that produced a set of equations of the applied pressures and shear stresses over the contact surface [42]. For the frictionless case, graphs of normal pressure distributions for progressive ratios of the half-sleeve length (b) to the cylinder radius (a) are plotted in Figure 7-16. For the case of complete adherence, the normal and shear contact stresses are presented in Figure 7-17.

In both cases, the highest stresses occur at the ends of the sleeves. A comparison of the normal stress plots shows that the graphs for the friction and adherence cases are very similar. This is to be expected because the effects of friction on normal stress are not significant. It can also be seen from the shear-stress distribution for the case of complete adherence (Fig. 7-17b) that the direction of the shearing stress at a specific location depends on the b/a ratio. Moreover, there is a value of b/a of approximately $\frac{1}{16}$ for which there is very little shearing stress except near the ends of the sleeve [43, p. 542]. Evidently, the geometry of the rigid sleeve could be modified to obtain a more uniform distribution of contact stress, thereby minimizing the edge effect stress concentration.

8

BUCKLING

The word *buckling* to many persons connotes a mode of failure. But there are a large number of examples where the buckling mechanism is used functionally, for example, in snap switches. This text, however, is mainly concerned with preventing buckling, since buckling causes nonuniform stressing and can lead to catastrophic failure of parts.

8-1 SHIFT FROM EFFICIENT TO INEFFICIENT STRESS PATTERN

Buckling usually begins with a type of stress pattern that is either compression, bending, torsion, or uniform shear. In the case of a column loaded in compression, for example, uniform stress exists. But when a critical load is reached, the column begins to deflect about the section axis of least moment of inertia, and the stress pattern changes from uniform compression to compression and bending. With a slight increase in load above the critical load, the bending stress exceeds the bending strength of the column material, and the column buckles.

The *critical load* can be defined as the maximum compressive load at which the state of the column changes from stable to unstable equilibrium. The stable column will continue to bend as the applied load approaches the critical load. Catastrophic failure takes place when the applied load exceeds the critical load.

The Euler critical load for the simply supported and centrally loaded column in Figure 8-1a can be determined by constructing the free-body diagram of the system as shown in Figure 8-1b. Following the development given in Ref. 34 [p. 395],

106

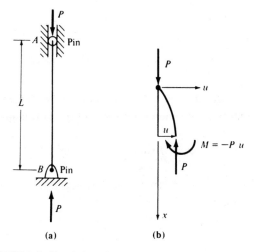

FIGURE 8-1. Fundamental case. [From Ref. 34, p. 393.]

the bending moment, M, in the column is given by

$$M = -Pu \qquad (8\text{-}1)$$

where P is the applied column load and u is the transverse deflection of the column. Another mathematical expression for the bending moment in the column can be determined with the assumption that the bending deflection u of the column is small. The bending moment M is related to the curvature (d^2u/dx^2) of the column by

$$M = EI\left(\frac{d^2u}{dx^2}\right) \qquad (8\text{-}2)$$

where E is the modulus of elasticity of the column material and I is the minimum moment of inertia of the column cross section.

Combining Eqs. (8-1) and (8-2) by eliminating M gives

$$-Pu = EI\left(\frac{d^2u}{dx^2}\right)$$

or

$$\left(\frac{d^2u}{dx^2}\right) + k^2u = 0 \qquad (8\text{-}3)$$

where

$$k^2 = \frac{P}{EI} \tag{8-4}$$

Equation (8-3) is the ordinary differential equation for the deflection of the column as a function of x and k. The general solution to Eq. (8-3) is given by

$$u = A \sin kx + B \cos kx \tag{8-5}$$

The constant coefficients A and B can be found by applying the boundary conditions for the column of length L:

At

$$x = 0, \quad u = 0 \tag{8-6}$$

At

$$x = L, \quad u = 0 \tag{8-7}$$

The boundary condition Eq. (8-6) can be used with Eq. (8-5) to yield

$$B = 0$$

Substitution of boundary condition Eq. (8-7) into Eq. (8-5) yields

$$0 = A \sin kL \tag{8-8}$$

The nontrivial solution for Eq. (8-8) requires that

$$kL = n\pi \quad (n = 0, \pm 1, \pm 2, \ldots) \tag{8-9}$$

Both sides of the Eq. (8-9) are squared to give

$$k^2 L^2 = n^2 \pi^2 \tag{8-10}$$

Substitution of the expression for k^2 from Eq. (8-5) into Eq. (8-10) yields the *Euler critical load*, P_c. For the fundamental case ($n = 1$), as shown in Figure 8-1a

$$P_c = \frac{\pi^2 EI}{L} \tag{8-11}$$

The deflection equation for the column is given by

$$u = A \sin\left[\left(\sqrt{\frac{P}{EI}}\right)x\right] \tag{8-12}$$

The critical stress, σ, can be found by dividing the critical load, P_c, by the cross-sectional area, A, of the column.

$$\sigma = \frac{P_c}{A} = \left[\frac{\pi^2 E}{L^2}\right]\frac{I}{A} \tag{8-13}$$

By defining

$$r^2 = \frac{I}{A} \tag{8-14}$$

where r is the *radius of gyration*, Eq. (8-13) becomes

$$\sigma = \frac{\pi^2 E}{(L/r)^2} \tag{8-15}$$

The expression (L/r) is known as the *slenderness ratio*.

The critical load for columns with end conditions other than the fundamental case can be determined by

$$P_c = \frac{\pi^2 EI}{L_e^2} \tag{8-16}$$

where L_e is defined as the effective length of the column. Effective lengths and the critical-load equations for various end conditions are presented in Table 8-1.

A shift from a stronger to a weaker stress pattern also can occur for the case of a long shaft subjected to torsion. As the torsion load on a long shaft is increased above the critical torque, the stress pattern shifts from torsion to bending, that is, to a less efficient stress pattern. The *critical torque*, T_c, can be defined as the torsional load at which the state of equilibrium of the shaft changes from stable to unstable.

To determine an analytical expression for the critical torque, we consider the simply supported shaft of uniform and arbitrary cross section in Figure 8-2a. The shaft is loaded at its ends with a torsional couple T. When the applied load reaches a critical value, the shaft will buckle and its geometric configuration can be ap-

TABLE 8-1. End Condition Effects on Critical Load [34, p. 408]

End Condition		Effective Length (L_e)	Critical Compressive Load (P_e)
Simply supported (pinned-pinned)		$L_e = L$	$P_e = \dfrac{\pi^2 EI}{L^2}$
Fixed-fixed		$L_e = \dfrac{L}{2}$	$P_e = \dfrac{4\pi^2 EI}{L^2}$
Free-fixed		$L_e = 2L$	$P_e = \dfrac{\pi^2 EI}{4L^2}$
Pinned-fixed		$L_e = 0.7L$	$P_e \cong 2.04\,\dfrac{\pi^2 EI}{L^2}$

proximated by a space curve as shown in Figure 8-2*b*. Following the development in Ref. 8 [pp. 292–294], the curve can be described by

$$y = g_1(x)$$
$$z = g_2(x)$$

$$(8\text{-}17)$$

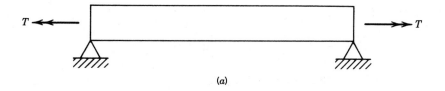

(a)

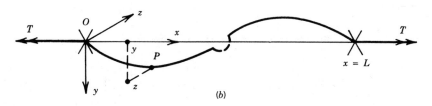

(b)

FIGURE 8-2. Buckling of shafts by torsion. [From Ref. 3, p. 294.]

A variable u can then be defined as a vector representation in the complex plane of the space deflection curve [8, p. 293]:

$$u = y + jz \tag{8-18}$$

The differential equation for torsion deflection is [8, p. 293]

$$\frac{d^2u}{dx^2} + \left[\frac{jT}{EI}\right]\frac{du}{dx} = 0 \tag{8-19}$$

The general solution to this first-order differential equation in terms of du/dx is

$$\frac{du}{dx} = A \cos\left(\frac{Tx}{EI}\right) - jA \sin\left(\frac{Tx}{EI}\right) \tag{8-20}$$

Integrating Eq. (8-20) yields

$$u = \left[\frac{AEI}{T}\right]\left[\sin\left(\frac{Tx}{EI}\right) + j\cos\left(\frac{Tx}{EI}\right) + B\right] \tag{8-21}$$

The constants A and B can be determined by applying boundary conditions.
   A first boundary condition at $x = 0$ requires that

$$y = z = 0$$

and therefore

$$u = y + jz = 0$$

Substitution of $u = 0$ into Eq. (8-21) yields

$$B = -j$$

Substitution of B into the deflection Eq. (8-21) yields

$$u = \left[\frac{AEI}{T}\right]\left\{\sin\left(\frac{Tx}{EI}\right) - j\left[1 - \cos\left(\frac{Tx}{EI}\right)\right]\right\} \tag{8-22}$$

The other boundary condition at $x = L$, where L equals the length of the bar requires that

$$y = z = 0$$

therefore

$$u = 0 \tag{8-23}$$

The nontrivial solution for this boundary condition requires that

$$\sin\left(\frac{TL}{EI}\right) - j\left[1 - \cos\left(\frac{TL}{EI}\right)\right] = 0 \tag{8-24}$$

or

$$\frac{TL}{EI} = 2n\pi \quad (n = 1, 2, 3, \ldots) \tag{8-25}$$

The lowest critical torque for a simply supported shaft of uniform cross section can then be determined as ($n = 1$)

$$T_c = 2\pi EI/L \tag{8-26}$$

The shape of the space deflection curve is given by

$$u = y + jz = C\left\{\sin\left(\frac{2\pi x}{L}\right) + j\left[1 - \cos\left(\frac{2\pi x}{L}\right)\right]\right\} \tag{8-27}$$

where

$$C = \frac{AEI}{T} \qquad (8\text{-}28a)$$

The maximum shear stress for a solid shaft in torsion can be expressed as [33, pp. 15–34]

$$\tau_{max} = \frac{16T}{\pi D^3} \qquad (8\text{-}28b)$$

where D is the diameter of the shaft.

The critical shear stress can be defined as the maximum allowable shear stress in the shaft prior to buckling. Substitution of Eq. (8-26) into Eq. (8-28b) yields an expression for the critical shear stress in a solid round shaft

$$\tau_c = \frac{16T_c}{\pi D^3} = \frac{32EI}{\pi D^3 L} \qquad (8\text{-}28c)$$

In a thin-walled tube of thickness t, the critical shear stress due to torsion is [27]

$$\tau_c = \frac{T_c}{2\pi r^2 t} \qquad (8\text{-}29)$$

where r is the outer radius of the tube.

8-2 BUCKLING IN MACHINE AND STRUCTURAL COMPONENTS

Consider the helical spring pictured in Figure 8-3. If the spring is too long or too slender and a load is applied, the spring will buckle. Figure 8-3 shows a buckling curve plotted on a graph of deflection divided by length versus length divided by the mean coil diameter. Buckling occurs for conditions above the line. This figure shows that for an l_0/D ratio of 4, a deflection of almost 20 percent of the spring length would be possible without buckling.

In the case where buckling does occur, a shift to a less efficient stress pattern takes place. The compression load on the helical spring produces a torsional stress distribution in the entire coil wire. But with buckling, the spring deflects in the middle, outward from the spring centerline, and bending stresses are induced with the stresses varying from zero at the ends of the spring to a maximum at the center of the spring.

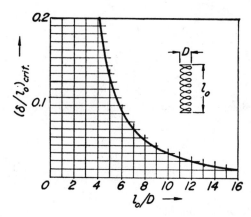

FIGURE 8-3. Buckling of a coil spring with hinged ends. [From Ref. 8, p. 260.]

A thin-walled tube loaded in compression will buckle locally (see Fig. 8-4) if the load exceeds a critical value. Again, during buckling there is a shift from a strong compressive-stress pattern to a compressive-stress pattern combined with a weak localized bending pattern. A similar strong-to-weak-stress pattern shift is found with a thin but deep beam subject to bending, a uniformly loaded arch, a thin-walled rectangular box, or an angle loaded in compression (see Figures 8-5, 8-6, 8-7, and 8-8).

The shift from a strong-to-weak-stress pattern is usually catastrophic. The uniform-stress distribution allows significant strain energy storage, but the shift to a nonuniform-stress pattern means that the stresses in certain areas must increase substantially since there is no other way to transfer or dissipate the stored strain energy. If the new stress pattern is considerably less efficient (in terms of strain energy storage capacity) than the original pattern, a serious failure will take place.

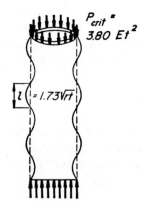

FIGURE 8-4. Buckling of rotationally symmetrical thin-walled tube in end compression. [From Ref. 8, p. 264.]

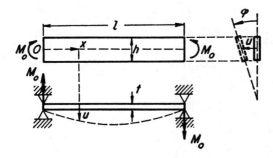

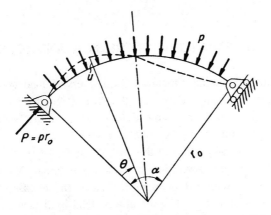

FIGURE 8-5. Thin rectangular beam subjected to bending moments at its ends in its stiff plane. [From Ref. 8, p. 283.]

FIGURE 8-6. Buckling of an arch, hinged at both ends, under uniform radial pressure p, with central angle α and radius r_0. [From Ref. 8, p. 281.]

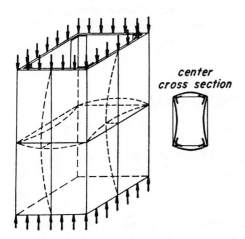

center
cross section

FIGURE 8-7. Buckling of box column made from four corner angles and four thin side plates under compression. [From Ref. 8, p. 302.]

115

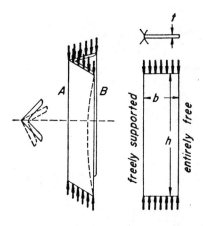

FIGURE 8-8. Buckling of the legs of an angle, hinged at its end under compressive load. [From Ref. 8, p. 305.]

8-3 CRITICAL LENGTHS FOR COMPRESSION AND TORSION

The equations for calculating critical axial compressive load and for computing critical torque can be used to determine the critical length of a member loaded respectively in compression or torsion. *Critical length* is the length of a member at which buckling occurs for a given load or torque.

Figure 8-9 gives a comparison of the critical lengths of solid and hollow steel shafts subjected to compressive force or torsional moment. For the hollow and solid shafts, the cross-sectional areas are the same. In the case of axial loading, the critical length is proportional to the critical load by a factor of $1/\sqrt{I}$. In the case of torsional loading, the critical length is inversely proportional to the critical load by a factor of $1/I$. For both compressive or torsion loading, the critical length of the hollow tube is significantly greater than that of the solid shaft. This is expected because the cross-sectional area of the hollow shaft exists away from the centroidal axis, resulting in a more efficient stress-carrying geometry.

8-4 DESIGNING TO PREVENT BUCKLING—EXAMPLES AND DISCUSSION

What can be done to prevent buckling? One design change that can be made is to add stiffeners. Figure 8-10 shows the use of a stiffener to prevent buckling in the top plate of a triangular brace. Figure 8-11 pictures the use of two stiffeners to prevent buckling in a web.

Figure 8-12 pictures a box connected with four spacer bolts to an angle iron frame. If the box is loaded downward, buckling will occur in the thin back of the

Geometry and Loading	System Schematic	Critical Load	Critical Length, L_c (ft)
Simply supported solid column in compression	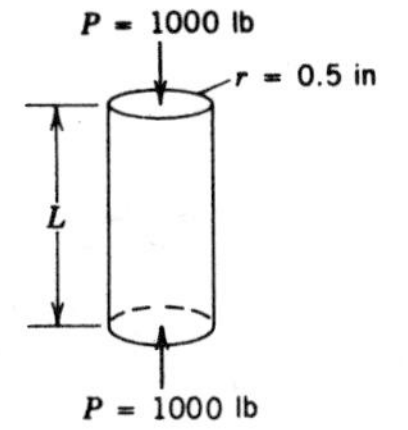 $P = 1000$ lb $r = 0.5$ in L $P = 1000$ lb	$P_c = \dfrac{\pi^2 EI}{L^2}$	10.04
Simply supported tubing in compression	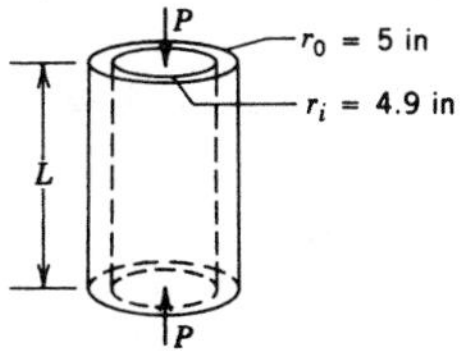 P $r_0 = 5$ in $r_i = 4.9$ in L P	$P_c = \dfrac{\pi^2 EI}{L^2}$	281.3
Simply supported solid column in torsion	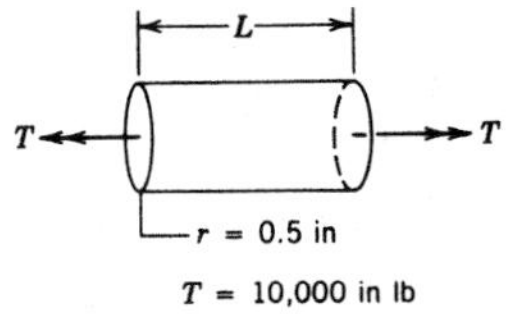 L T ← → T $r = 0.5$ in $T = 10,000$ in lb	$T_c = \dfrac{2\pi EI}{L}$	77.11
Simply supported tubing in torsion	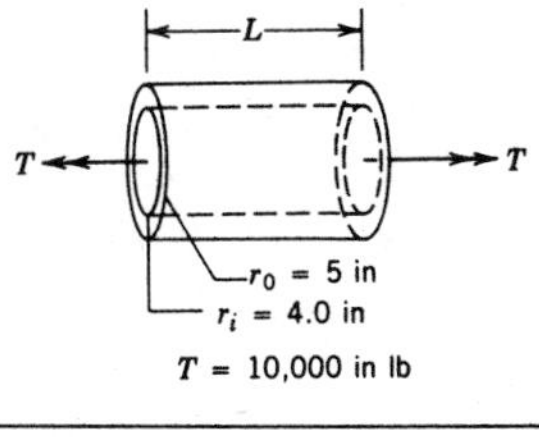 L T ← → T $r_0 = 5$ in $r_i = 4.0$ in $T = 10,000$ in lb	$T_c = \dfrac{2\pi EI}{L}$	60,451

FIGURE 8-9. Critical length comparison of steel columns.

box. By adding channels that also act as spacers, the box is stiffened and an improved design results.

There are other schemes that can be used to prevent buckling. One of the best is to use diagonal bracing—the tetrahedron-triangle principle. The channel in Figures 8-13a and b can warp when subjected to torsion. The use of diagonal bracing in Figure 8-13c prevents this from happening.

The relative effect of diagonal bracing versus rectangular bracing, or no bracing at all, can be seen in Figure 8-14, where a box-shaped member is fixed on the far end and torsionally loaded on the pinned end. The diagonal bracing stiffens the container and helps it resist a torsional load. The bracing prevents bending or

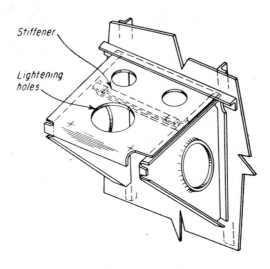

FIGURE 8-10 Stiffener provided to allow thinner and lighter shelf sheet. [From Ref. 9, p. 109.]

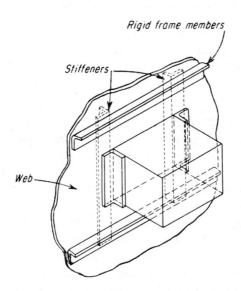

FIGURE 8-11. Stiffeners used to carry moment loads to rigid frame and prevent unsupported web from "oilcanning." [From Ref. 9, p. 109.]

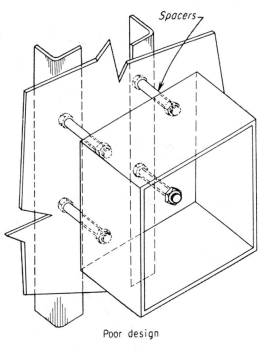

Poor design

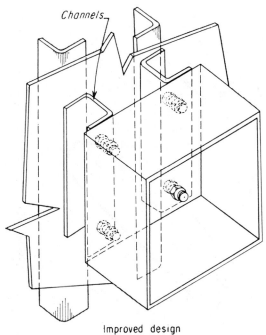

Improved design

FIGURE 8-12. Channels employed in improved design to best space box from attachment wall to prevent bending from side loads. [From Ref. 9, p. 109.]

119

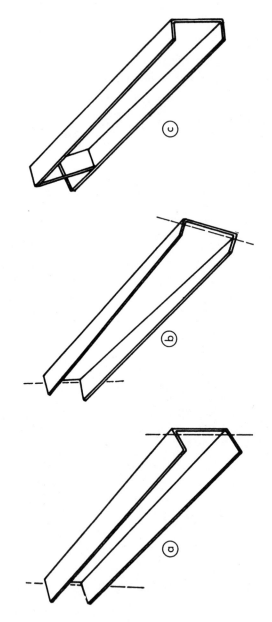

FIGURE 8-13. Diagonal bracing used to make rigid end connections. [From Ref. 2, p. 3.6-15.]

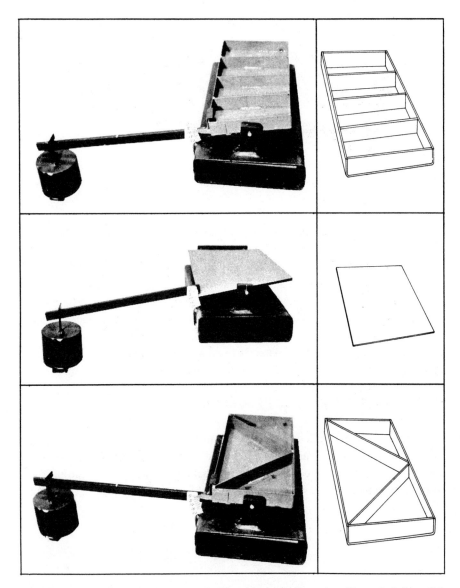

FIGURE 8-14. Relative effectiveness of different types of bracing. [From Ref. 2, p. 3.6-12.]

buckling from taking place. Clearly, diagonal bracing is much better than rectangular bracing.

Figure 8-15 shows a second comparison of rectangular versus diagonal bracing. The boxes pictured are loaded on one corner and supported on the other three corners. It is clear that the diagonal bracing produces the stiffest part.

FIGURE 8-15. Diagonal bracing used to prevent frame from twisting. [From Ref. 2, p. 4.3-8.]

Figure 8-16 shows a curved I-beam with a vertical load applied at the horizontal end of the beam. This load could cause the web to buckle. The addition of a stiffener would prevent buckling. The stiffener helps transfer the stress from the web into the top and bottom flanges of the beam.

Examples have been given where buckling takes place with initial stress patterns of compression, uniform shear, torsion, and bending. Notice that buckling never occurs with tensile loading and that the addition of stiffeners, in effect, prevents the shift to less efficient stress patterns. Buckling can in most cases be prevented by recognizing and preventing a shift from strong-to-weak-stress patterns.

8-5 DESIGNING LIGHTWEIGHT FRAMES

Hollow tubes or thin-walled cylinders are often used in designs that require high strength and minimum weight. However, when a tubular member is loaded in compression, failure can occur by column buckling, yielding, or local buckling. The specific mode of failure is determined by the cross-sectional geometry and the

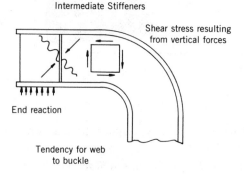

FIGURE 8-16. Stiffener used to prevent buckling in web section where high compressive loads might exist. [From Ref. 2, p. 4.3-5.]

length of the tube. To achieve a maximum strength-to-weight ratio in tubular frames, we need to optimize the length of the tubing for a given wall thickness.

With tubing subjected to simple uniaxial compressive loading, the maximum possible length is where two modes of failure are likely to occur simultaneously. These modes are column buckling and yielding, yielding and local buckling, or column buckling and local buckling. The optimum tube length can be determined by first calculating a theoretical value for the ratio of the inner to the outer radii of the tubing from

$$(r_i/r_o)_t = \frac{(2/\sqrt{3})[E/(1 - \mu^2)^{1/2}] - S_y}{S_y + 2/\sqrt{3}[E/(1 - \mu^2)^{1/2}]} \tag{8-30}$$

where

E = elastic modulus

S_y = compressive yield strength of the column material

μ = Poisson's ratio

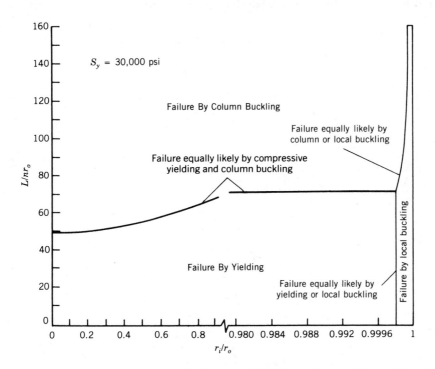

FIGURE 8-17. Failure modes for steel tubing. [Reprinted from *Machine Design*, May 15, 1975, p. 88. Copyright 1975 by Penton/IPC, Inc., Cleveland, Ohio.]

The value of $(r_i/r_o)_t$ from Eq. (8-30) can be compared to the ratio of radii (r_i/r_o) for the actual tubing. If the theoretical radii ratio $(r_i/r_o)_t$ is greater than the actual radii ratio (r_i/r_o), then

$$\frac{L}{nr_o} = \left[\frac{\pi}{2}\right]\left[\frac{E\left(1 + (r_i/r_o)^2\right)}{S_y}\right]^{1/2} \qquad (8\text{-}31)$$

If $(r_i/r_o)_t$ is less than (r_i/r_o) then

$$\frac{L}{nr_o} = \left[\frac{\pi}{2}\right]\left[\left(\frac{3^{1/2}(1-\mu^2)}{2}\right)^{1/2}\right]\left[\left(1 + \frac{r_i}{r_o}\right)\left(\frac{1 + (r_i/r_o)^2}{(1 - r_i/r_o)}\right)\right]^{1/2}$$

where n is defined as the end condition coefficient. Information on determining n can be found in Roark [30, p. 259]. If $(r_i/r_o)_t$ equals (r_i/r_o), either equation for L/nr_o can be used.

Having determined L/nr_o, we can find the optimum tube length that yields the most efficient stress pattern at the maximum strength-to-weight ratio. With reference to Figure 8-17, the optimum tubing lengths are at the boundaries of yielding and column buckling or column buckling and local buckling.

9

IMPACT

Impact occurs in machines or structures when a load is applied for a duration that is less than one half the fundamental period of the impacted member. A load that is applied for a duration longer than three times the fundamental period of an impacted member or structure is defined as a static load.

When members or structures are loaded, they deflect and store energy internally. The energy storage capacity of an element in tension or compression is proportional to the stress squared divided by the modulus of elasticity of the element. The total energy storage capacity of a body is equal to the sum of the capacities of all the elements. Hence energy capacity or impact capacity is improved by more uniform-stress patterns with each element in a body stressed to its maximum, assuming each element has the same strength. In other words, for improved energy capacity or impact capacity, design for uniform stress: uniform tension, compression, shear, and bearing.

9-1 ANALYSES OF AXIAL AND TORSIONAL IMPACT

Consider Figure 9-1 which depicts an axially loaded rod *BC* and a torsionally loaded shaft *EF*. The following analyses for axial and torsional impact are made under the following assumptions [34]:

1. The elastic behavior of each of the impacted members is linear.

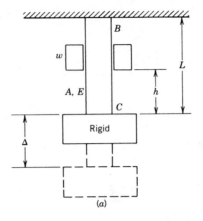

(a)

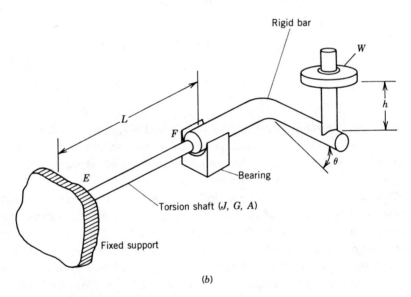

Rigid bar

W

h

L

F

E

θ

Bearing

Torsion shaft (J, G, A)

Fixed support

(b)

FIGURE 9-1. Impact loadings. [From Ref. 34, p. 618.]

2. All the energy of the impact force (weight W) is transformed into elastic strain energy.

3. Maximum stresses and deflections occur instantaneously in the impacted bodies.

4. Rigid elements in the body do not absorb energy.

5. The impact stresses are large compared to the static load stresses due to the weight of the bodies.

6. The work done by the weight of the deformed bodies is negligible.

It should be cautioned that the assumptions made may not always yield accurate results in actual practice.

For both cases shown in Figure 9-1, the maximum deflection of the impacted body is Δ. The maximum deflection in (a) for the axially loaded rod can be described in terms of the impact kinetic energy U as

$$U = \tfrac{1}{2} mv^2 = \frac{Wv^2}{2g}$$

where v is the impact velocity of the weight W. The impact deflection is given by

$$\Delta = \left[\frac{2U\Delta_{st}}{W}\right]^{1/2} = \left[\frac{2U}{k}\right]^{1/2}$$

where Δ_{st} is the static deflection of the body due to weight W and k is the effective spring rate (stiffness) of the rod. An *equivalent static force* that would initiate the impacted deflection Δ is written as

$$F_e = W\left[\frac{v^2 k}{Wg}\right]^{1/2} = [2Uk]^{1/2} \tag{9-1}$$

The impact stress can be defined as

$$\sigma = \frac{F_e}{A}$$

where

$$k = \frac{AE}{L}$$

Substituting the expressions for σ and k into Eq. (9-1) yields the impact stress

$$\sigma = \left[\frac{2UE}{AL}\right]^{1/2} = \left[\frac{2UE}{V}\right]^{1/2}$$

where $V = AL$ is the volume of the rod.

The energy absorbed by the rod upon impact is therefore

$$U = \frac{\sigma^2 V}{2E}$$

The equations of deflection, equivalent load, and stress for *torsional impact* are analogous in form to those for axial impact loadings. By referring to Figure 9-1*b*, we can write the maximum angular deflection θ of the solid shaft in radians as

$$\theta = \left[\frac{2U}{k_t}\right]^{1/2}$$

where k_t is the torsional spring rate.

The *equivalent static torque* required to initiate this deflection is

$$T_e = [2Uk_t]^{1/2}$$

The shear-stress equation for a torsionally impacted solid round shaft is

$$\tau = 2\left[\frac{UG}{AL}\right]^{1/2} = 2\left[\frac{UG}{V}\right]^{1/2}$$

Analogous to the axial case, the energy stored in the rod upon torsional impact can be written as

$$U = \frac{\tau^2 V}{2G}$$

9-2   ENERGY STORAGE FOR EFFICIENT VERSUS INEFFICIENT STRESS PATTERNS

Figure 9-2*a* shows a tensile bar that has three separate segments. The center segment has a smaller area than the other two segments. The load F produces the stress diagram in Figure 9-2a. Where the cross-sectional area is the smallest, clearly the stress is the highest. A bar with a constant cross-sectional area would have uniform stress throughout. The maximum stress is the same in both bars shown in Figure 9-2. Evidently, the uniformly stressed bar would store more energy for the same maximum stress. Tensile bar (*a*) has a portion having a larger area which is not being stressed to a maximum amount. Since energy storage

$$U \cong \sum_{i=1}^{n} \left(\frac{s_i^2}{2E}\right)V_i$$

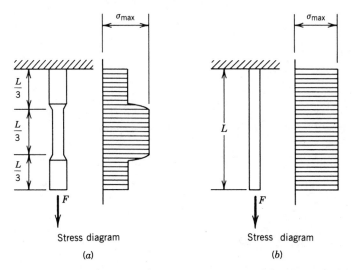

Stress diagram

(a)

Stress diagram

(b)

FIGURE 9-2. Two tensile bars having the same strength but different energy-absorbing abilities under load. [From Ref. 2, p. 3.1-6.]

where

V_i = volume of element i

s_i = stress in element i

E = modulus of elasticity of the element

the bar (b) would be able to store more energy than (a).

For example, following Juvinall [4, pp. 180–181], we consider two different round rods subjected to tensile impacts as shown in Figure 9-3. The energy storage capacity of the uniform rod is 37.5 percent greater than that of the stepped rod.

The stored energy in each rod is given by

$$U = \frac{\sigma^2 V}{2E}$$

For the bottom half of the stepped rod, let

$$\sigma = S_y \qquad V = V_a$$

The energy stored in the bottom half is then given by

$$U = \frac{S_y^2 V_a}{2E}$$

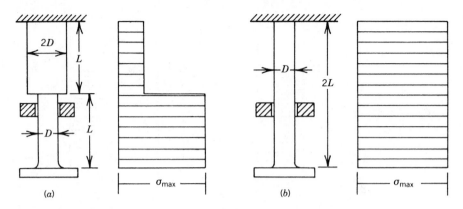

FIGURE 9-3. Impact effects on nonuniform cross sections. [From Robert C. Juvinall, *Engineering Considerations of Stress, Strain, and Strength*, McGraw-Hill, New York, 1967, p. 181. Reproduced by permission.]

The top half of the stepped rod has four times the volume and one fourth the stress of the bottom half. The energy stored in the top half is therefore

$$U = \frac{S_y^2 V_a}{8E}$$

The total energy capacity for the stepped rod is the sum of its parts and can be written as

$$U_a = \frac{5 S_y^2 V_a}{8E}$$

For the uniform rod, the energy storage capacity is the same for the upper and lower halves of the rod and is given by

$$U_b = \frac{S_y^2 V_a}{E}$$

Therefore, as previously stated, the energy storage capacity of the uniform rod is 37.5 percent greater than that of the stepped rod. Also, on a energy storage to weight basis, the uniform rod is four times more efficient than the stepped rod. This is to be expected because of the nonuniform distribution of stress in the stepped rod.

As an illustration of the effect of stress concentration on energy storage, consider the notched and unnotched tensile bars pictured in Figure 9-4. Except for the notched portion, both bars have constant cross-sectional areas with length. The uniform sections have a uniform stress diagram. Since energy storage is propor-

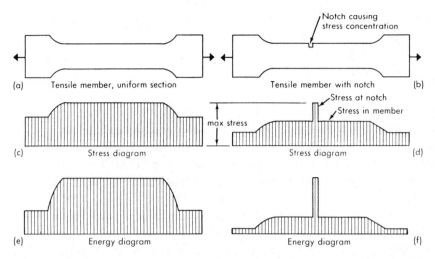

FIGURE 9-4. Effect of notch on energy-absorbing capacity. [From Ref. 2, p. 3.1-7.]

tional to the stress squared, where the stress diagram is uniform, the energy diagram is uniform. But in the notched member, the notch causes a rise in stress, and the maximum stress occurs at the notch. So there are portions on both the left- and the right-hand side of the notch that carry a lower amount of stress (because if the member carries a higher amount of stress, then failure would occur at the notch). The energy diagram shows that the notched member does not store nearly as much energy as the uniformly stressed member. In this example, and in general, for impact, a designer should avoid notches and other stress concentrations. High-impact strength parts should be designed for uniform stress. Therefore all corollaries such as the tetrahedral-triangle principle, the mating-surface principle, and the hollow-shaft principle are applicable.

9-3 PREDICTING THE BEHAVIOR OF MEMBERS UNDER IMPACT LOADING

In design, it is sometimes necessary to predict the behavior of components and structures under impact loading. Through extensive experimentation, empirical stress-impact factors have been determined for various applications. The *stress-impact factor* is the factor by which a static load and deflection are increased due to impact loads.

For the general case of impact loading, consider a spring that is axially impacted by a weight W released from a height h. In terms of the static deflection Δ_{st}, the maximum dynamic deflection of the spring upon impact is

$$\Delta = \Delta_{st} \left(1 + \left[1 + \left(\frac{2h}{\Delta_{st}} \right) \right]^{1/2} \right)$$

and the *equivalent static force* required to deflect the spring Δ is given by

$$F_e = W\left(1 + \left[1 + \left(\frac{2h}{\Delta_{st}}\right)\right]^{1/2}\right)$$

The expression $(1 + [1 + (2h/\Delta_{st})]^{1/2})$ is defined as the analytical stress factor. If experimental stress factors are not available for a particular application, the analytical stress factor can be used. However, the stress factor of the real system may be much higher than the analytical result [4, p. 170]. The final factor of safety for the system must be determined by the designer.

As a specific example, consider the rectangular steel beam pictured in Figure 9-5. The beam is simply supported and subjected to an impact load due to a weight W released from a height h. The stress-impact factor, the maximum deflection, and the equivalent static force of the beam can be determined as follows [4, p. 171].

The static deflection of the beam is given by

$$\Delta_{st} = \frac{PL^3}{48EI} = \frac{50(30)^3}{48(30)(10^6)(1/12)(1)(3)^3}$$

$$= 4.17 \times 10^{-4} \text{ in}$$

The impact factor, *IF*, is therefore

$$IF = 1 + \left[1 + \left(\frac{2h}{\Delta_{st}}\right)\right]^{1/2} = 1 + \left[1 + \left(\frac{2(2)}{\Delta_{st}}\right)\right]^{1/2}$$

$$= 98.985$$

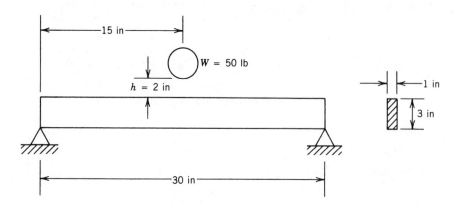

FIGURE 9-5. Impact-loaded beam.

where $h = 2$ in. The maximum deflection is given by

$$\Delta = \Delta_{st}(IF) = 0.0412 \text{ in}$$

and the equivalent static force is found to be

$$F_e = W(IF) = 4950 \text{ lbf}$$

The equations for calculating the stress-impact factor, maximum dynamic deflection, and the equivalent static force can also be used to compare the effects of axial and transverse impact loads on a solid, round bar. For equivalent impact loadings, Table 9-1 lists impact deflections and equivalent static loads for a specific cantilever beam and column. The analysis assumed that column buckling did not occur. Referring to Table 9-1, we see that the impact deflection of the cantilever beam is 23 times greater than the deflection of the column. This is expected because a long bar is less rigid and more susceptible to bending in the transverse direction than in the axial direction. The equivalent static force of the column is 23 times greater than the force on the cantilever. The normal stress on the cantilever beam is approximately 3.6 times greater than that of the column.

9-4 EXAMPLES AND DISCUSSION OF IMPROVED DESIGNS

The effect of a stress concentration can be seen by inspection of isochromatic patterns for a stepped part subjected to an impact blow on the left end. Figures 9-6 and 9-7 show a traveling stress wave and a concentration of fringes at the step. To the right of the step, the stress wave expands because of the larger cross-sectional area, and the stress is reduced. The lowest stress would be found in the area on the right, whereas the highest stress would exist just to the left of the step in the portion with the smallest cross-sectional area. This stepped-shaped member would be inefficient as an impact-absorbing device because all material is not being stressed uniformly.

Consider the problem of comparing the impact capacity of a uniform rod of diameter D with the rod shown in Figure 9-8. An analysis would predict, neglecting the effect of the stress concentration in the notch, that the uniform rod would absorb 16 times more energy than the notched rod. A similar analysis would show that removal of the bolt material pictured in Figure 9-9, would produce a stronger impact member. With the bolt, the maximum stress will occur in the threads where the stress concentrations exist. Thus the material on the bolt body is not being stressed to its maximum. Since the energy capacity is proportional to the stress squared times the volume that is being stressed, and since the bolt body is a large volume,

TABLE 9-1. Effect of Bar Configurations

Condition(s)	Schematic	Δ_{st}	IF	Δ (in)	F_e (lbf)	Normal Stress (psi)
Cantilever beam $W = 10$ lbf $h = 10$ in $d = 1$ in $E = 30 \times 10^6$ psi		$\dfrac{WL^3}{3EI}$	95.0	215×10^{-3}	950	9.7×10^4
Column $W = 10$ lbf $h = 10$ in $d = 1$ in $E = 30 \times 10^6$ psi		$\dfrac{WL}{AE}$	2171.8	9.2×10^{-3}	21,718	2.7×10^4

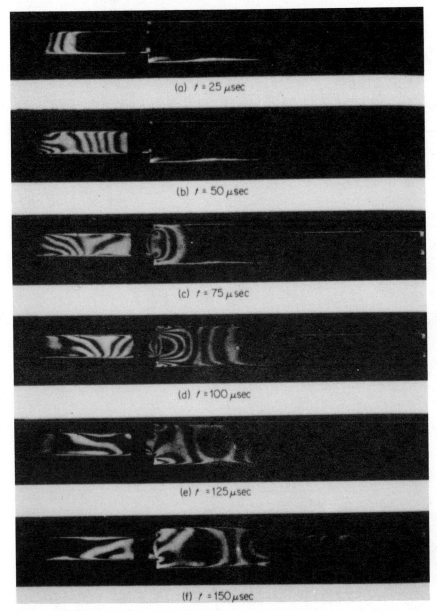

FIGURE 9-6. Isochromatic patterns in a bar with an enlarged central section at different instants after impact at the left-hand end, from $t = 0$ to 150μ sec. [From Ref. 1, p. 488.]

it should be stressed to its maximum. By removing material, we can store more impact energy in the bolt.

Figure 9-10 shows a bolt designed for impact. This would be an expensive bolt. But if bolt failure was a problem, this bolt could be employed. The shank of the

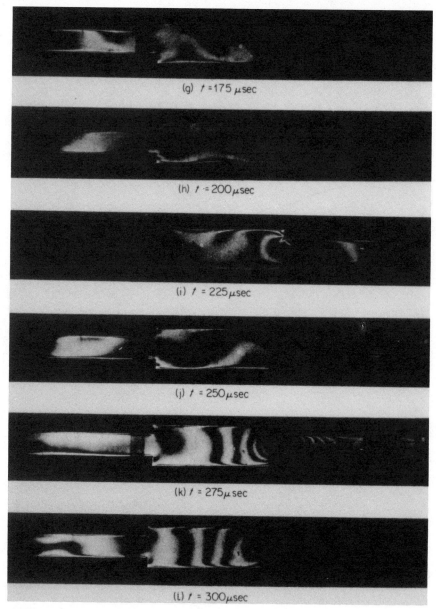

FIGURE 9-7. Isochromatic patterns in a bar with an enlarged central section at different instants after impact at the left-hand end, from $t = 175$ to $300\ \mu$ sec. [From Ref. 1, p. 489.]

bolt is reduced to the core of the threads to obtain high resilience, so that it can absorb as much energy as possible. Notice that smooth transition sections are used to reduce stress concentrations.

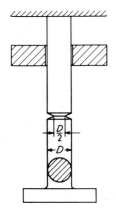

FIGURE 9-8. Notched circular tensile-impact bar. [From Robert C. Juvinall, *Engineering Considerations of Stress, Strain, and Strength*, McGraw-Hill, New York, 1967, p. 183. Reproduced by permission.]

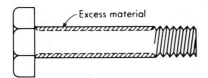

FIGURE 9-9. Removal of excess material from bolt to increase impact strength. [From Robert C. Juvinall, *Engineering Considerations of Stress, Strain, and Strength*, McGraw-Hill, New York, 1967, p. 184. Reproduced by permission.]

Shapes that are best for impact are those stressed uniformly during load. For tension or compression a constant cross-sectional area would produce the best results. For bending, a tapered beam, for example, a leaf spring, would be a good choice. For torsion, a thin-walled tube should be used.

9-5 DESIGN CONSIDERATIONS FOR MINIMIZING THE EFFECTS OF IMPACT

There are several design guidelines for reducing the potentially catastrophic effects of impact loadings on structures. These considerations, listed in Ref. 33, [p. 16–3], are as follows:

1. Minimize the loading speed of impact.
2. Reduce the mass of the impacting bodies.
3. Design for flexibility in the vicinity of the point of impact.
4. Stress the maximum possible volume of materials uniformly to their peak stress for maximum energy storage capacity.

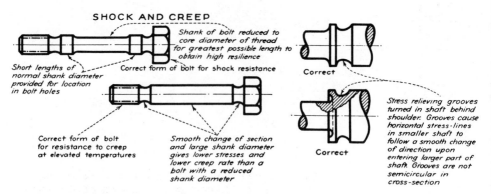

SHOCK AND CREEP

Shank of bolt reduced to core diameter of thread for greatest possible length to obtain high resilience

Short lengths of normal shank diameter provided for location in bolt holes

Correct form of bolt for shock resistance

Correct

Correct form of bolt for resistance to creep at elevated temperatures

Smooth change of section and large shank diameter gives lower stresses and lower creep rate than a bolt with a reduced shank diameter

Correct

Stress relieving grooves turned in shaft behind shoulder. Grooves cause horizontal stress-lines in smaller shaft to follow a smooth change of direction upon entering larger part of shaft. Grooves are not semicircular in cross-section

FIGURE 9-10. Bolt design modifications for shock and creep loadings. [From Ref. 10, p. 302.]

5. Avoid material and geometry factors that increase the local stress concentration of the impacting bodies. This can be accomplished by
   a. Using a ductile material with some capability for deformation to counter stress concentration effects.
   b. Avoiding sharp discontinuities, surface irregularities, and inhomogeneous materials in the design and manufacture of parts.

10

STRENGTH/DESIGN
OF JOINT ELEMENTS

A machine part is comprised of body parts and joint elements. In the previous portion of this text, the discussion centered mainly on designing body parts for uniform stress. In this chapter, the discussion deals with designing joint elements.

10-1 JOINTS

A *joint* is defined as a break in the continuity of a machine structure. Examples of joints include members that are riveted, welded, splined, pinned, taper pinned, screw fastened, and interested. Figure 10-1 pictures many of these joint types.

Each joint consists of body portions and joint elements. Figure 10-2 shows a simple butt-welded joint with body parts and joint elements identified. All joint elements, even good welds, should be recognized as breaks in the continuity of machine parts.

10-2 THE NEED FOR JOINTS

Joints provide a means for assembling machine members and also for transmitting forces or torques between machine parts. A joint, being a break in the continuity of a machine part, forces the flow of force to detour, often through circuitous and restricted paths, around the discontinuity. Joints are therefore inherently weak and

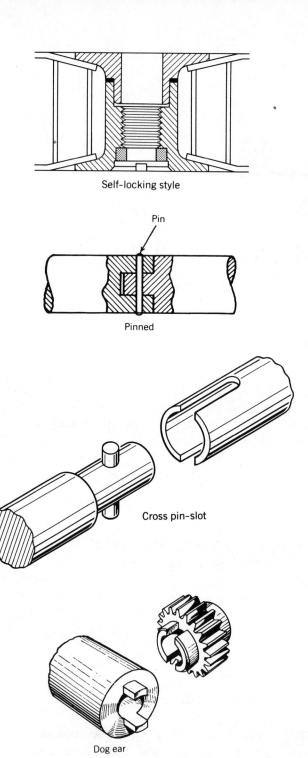

Self-locking style

Pin

Pinned

Cross pin–slot

Dog ear

FIGURE 10-1. Machine and structural joints. [From Ref. 12, p. 40; 13, p. 100; 14, p. 97; 15, p. 28-3, 28-5.]

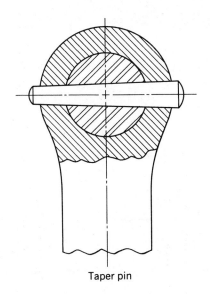

Taper pin

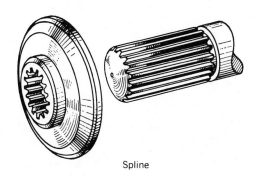

Spline

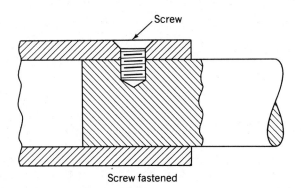

Screw

Screw fastened

FIGURE 10-1. (*Continued*)

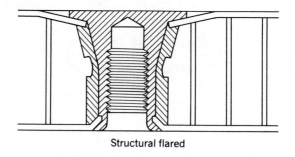

Structural flared

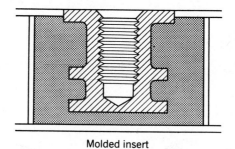

Molded insert

FIGURE 10-1. (*Continued*)

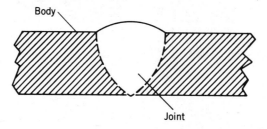

FIGURE 10-2. Single-vee butt weld joint. [From Ref. 16, p. 355.]

have to be reinforced in some manner. This reinforcement adds weight and explains why joints are generally inefficient.

10-3 THE NATURE AND EFFECT OF JOINTS

Consider a threaded rod with nuts on each end, with a load being applied to each nut. Figure 10-3 shows a photograph of the isochromatic pattern with fringe numbers for the bolt and nuts. The highest fringe number is 7.5, and this is found in the threaded portion near the bottom nut. The fringe number for the bolt portion

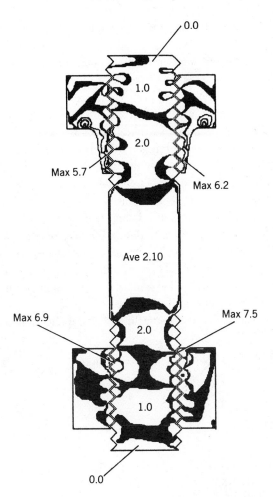

FIGURE 10-3. Isochromatic pattern for a tensioned threaded rod with conventional and tapered lip nuts.

midway between the nuts is 2.10. The fringe number is proportional to the shear stress with a higher fringe number indicating a higher shear stress.

Inspection of the two nuts will reveal the effect of nut shape on stress. The tapered nut on the top has a maximum fringe number of 6.2, whereas the bottom nut as mentioned has a maximum fringe number of 7.5. The tapered portion of the top nut clearly allows the stress to flow more gradually from the nut to the bolt than does the untapered bottom nut.

Note also that the first engaged thread of the bolt sees the highest stress. The second thread is less stressed, and so on for the third thread. The tapered portion of the nut reduces the magnitude of the stress on the first thread by using a less stiff (thinner) portion that is adjacent to the first engaged thread. This, matching of rigidities, is called elastic matching, and this principle will be discussed in Chapter 12.

Consider now the flow of force from the top nut, through the bolt, and into the bottom nut. What path does the flow take? The concentrated load P is applied as shown in Figure 10-4 on the tapered nut. The force flows from the point of load application into the nut and then through the thread elements into the bolt. The contact areas of the mated threads act like bridges between the nut and the bolt, and the force flows through these bridges from the nut into the bolt body. In the case of a threaded joint, there is a large number of bridges. Once the flow reaches the bolt body, it travels down to the bottom nut, first passing through the bridges formed by the thread contact areas. The density of the flow lines is greater at point 2 than at point 1. Hence the stress at 2 is higher than at 1.

The joint forces the flow of force to divide and pass over discrete concentrated bridges, which in this example are the thread contact areas. The force flow then rejoins to form a uniform-stress field on the other side of the joint, that is, in the bolt body. The force flow division can induce stress patterns of bending, torsion, and spot contact in the detour bridge pattern.

If a uniformly distributed load could be directly applied to each bolt end seg-

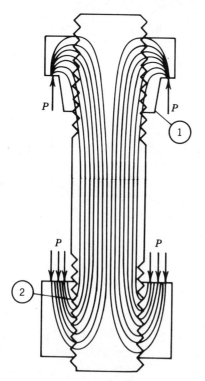

FIGURE 10-4. Stress flow through the tapered nut and threaded rod.

ment, a uniform stress would exist throughout the bolt. But because the load is transferred from the tapered nut through a joint, into the bolt, and through the bolt, and into the bottom nut through another joint, nonuniform-stress patterns of spot contact and bending are induced; and because of the development of these inefficient patterns, the highest strength-to-weight ratio assembly is not produced.

As a second example for the effect of joints, consider the force flow through a clevis joint, where a load F is applied at each end (see Figure 10-5). How does the force flow? Using Figure 10-6 and assuming a path from fork to blade, we find

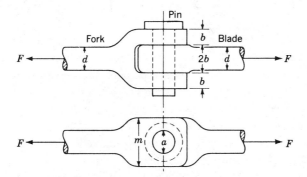

FIGURE 10-5. Yoke joint loaded in tension. [From Robert C. Juvinall, *Engineering Considerations of Stress, Strain, and Strength*, McGraw-Hill, New York, 1967, p. 12. Reproduced by permission.]

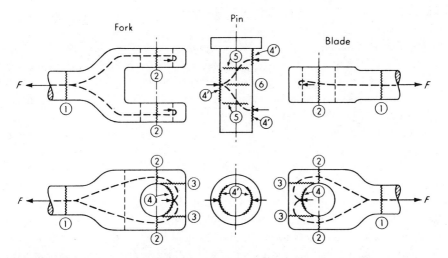

FIGURE 10-6. Flow of force and critical sections in yoke joint. [From Robert C. Juvinall, *Engineering Considerations of Stress, Strain, and Strength*, McGraw-Hill, New York, 1967, p. 12. Reproduced by permission.]

that the flow of force separates at 1 into the tangs of the fork and flows to the fork at 4, then into the pin at 4, through the pin to 4′ on the opposite side, into the blade at 4, and through and out of the blade at 1.

At section 1 in the fork, before the flow separates, uniform stress exists. But then the joint forces the flow to divide. Spot contact is found between the pin/fork and pin/blade surfaces at the interface of 4 and 4′. Bending stresses are found in the pin. Because of the nonuniform distribution of stress, material is wasted as portions of the joint are not stressed to their maximum strength. Force-flow visualization can therefore be used in the process of streamlining the joint by eliminating excess material or it can be used to identify critical areas where failure in the yoke may occur.

Figure 10-7 shows the stress flow path through a lap joint loaded on the left plate to the left and loaded on the top and bottom strap to the right. The straps are riveted together. The stress flows from the left plate into the rivets on the left. The flow is then carried by the rivets, to the top and bottom straps that carry the flow to the rivets on the right. The rivets then transfer the stress flow into the plate on the right.

If the left plate could be joined directly to the right plate, for example, by welding, both the rivets and the top and bottom strap would not be needed and a much more efficient joint would be obtained. The next section gives several principles that are useful for effecting more efficient joints.

10-4 PRINCIPLES USEFUL FOR JOINT DESIGN

What are four principles that are useful for joint design? *First*, use many bridges. For example, clothing designers use zippers rather than buttons where compact joints are needed. With numerous bridges, each bridge helps transmit the load between members, and the flow becomes more uniform since detours will not have to be made.

A door with three hinges has only three bridges, whereas the piano hinge on the door of a medicine cabinet offers many paths for the flow of force. These are trade-offs such as cost to consider, but for strength, the joint with the most bridges will be the most efficient.

The *second* principle that is useful for joint design is to make the bridges as short as possible. The longer the bridge is, the more material is needed. Two steel plates loaded in tension and to be fastened at their ends, could best be joined by butt welding. If the ends were overlapped and welded, material would be wasted; the bridges would be longer and the joint heavier. So, keep the bridges as short as possible.

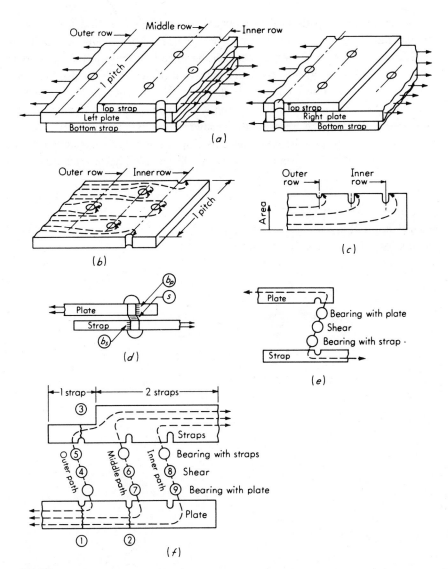

FIGURE 10-7. Flow of force through a triple-riveted butt joint. (*a*) Free-body diagram of sectioned joint; (*b*) force flow through plate to rivets; (*c*) diagram of force flow versus cross-sectional area; (*d*) force flow through rivet; (*e*) diagrammatic representation of force flow through rivet; (*f*) complete diagrammatic representation of force flow. [From Robert C. Juvinall, *Engineering Considerations of Stress, Strain, and Strength*, McGraw-Hill, New York, 1967, p. 14. Reproduced by permission.]

The *third* principle is to use uniform-stress tension, compression, or uniform shear in the bridges as much as possible. Try to avoid inefficient stress patterns like bending.

Fourth, shape the parts in the vicinity of the bridges in such a manner as to separate the force flows as gradually as possible in order to maintain a uniform-stress distribution. In other words, channel the force flow uniformly into the joint. Look, for example, at towers that carry high-voltage lines and visualize the force flow from the wire tower connection down to the tower base. Here structural members gather the tower weight and the wire load and channel it right into the four bolts that fasten the tower to the ground. There is an attempt to make the force flow gradually into the joint elements. The entire load is transmitted to the joint elements in such a manner as to stress the joint elements as uniformly as possible.

In review, the principles are to (1) use many bridges, (2) make the bridges as short as possible, (3) use uniform-stress patterns, and (4) gather the stress and channel it gradually into the joint elements. The next chapter presents several methods of achieving highly efficient joints.

11

METHODS OF ACHIEVING HIGHLY EFFICIENT JOINTS

In the previous chapter we listed principles that can be used to improve joint design. In this chapter, we present several methods for designing more efficient joints. The efficiency of a joint can be expressed as the ratio of the joint strength to the surrounding member strength. Several examples of efficient and inefficient joints are given.

11-1 THREE METHODS TO IMPROVE JOINT EFFICIENCY

There are three ways in which joint efficiency can be improved.

1. Use a large number of small fasteners widely and evenly distributed over a joint, leaving the major body shape and body stress patterns largely undisturbed.

2. Provide intermediate stress-gathering channels connecting the major body shape to a few large fasteners.

3. Use a combination of the two preceding methods.

11-2 EXAMPLES AND DISCUSSION

Figure 11-1 shows two pipe joint connections where a large number of small fasteners are employed to provide a good seal and an efficient joint. Imagine the

149

FIGURE 11-1. Two bolted pipe joint connections.

difference in bolt size and bearing-stress distribution if only four bolts were used to connect the flanges.

Another example is embodied in Figure 11-2. The compressor end plate contains many small bolts that are uniformly spaced. The legs supporting the compressor are bolted to the floor and act as force-gathering channels that transfer the compressor weight and any other loads to the floor.

Figure 11-3 shows five small V-belts employed in a pulley drive. This method of power transmission uses five small bridges rather than one large bridge. With other considerations being equal, such as cost, from the viewpoint of force flow, it would be better to use multiple V-belts rather than one large belt.

Figure 11-4a shows a four-bolt hole flange welded to a large pipe. It would be difficult for this assembly to transmit a bending moment. Force would have to flow from the pipe through the highly stress sensitive flange pipe joint into the flange

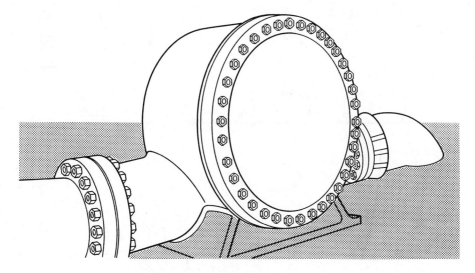

FIGURE 11-2. Compressor end plate.

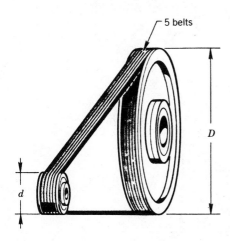

FIGURE 11-3. Multiple V-belt drive. [From Ref. 17, p. 117.]

and to the four bolts. But the force would not be gathered or channeled gradually into the four bolts and would have to pass through the highly stress sensitive pipe flange joint.

An improved design is shown in Figure 11-4b. Eight stress gatherers transmit the stress flow from the pipe gradually into the flange. This flange contains eight-bolt holes rather than four, applying the suggestion to employ a large number of fasteners widely and evenly distributed. Figures 11-4c, d, e, and f reveal additional cases where stress gatherers are used in joint design.

Figure 11-5 shows two pedestal bases. Bracket (b) will be stronger in carrying

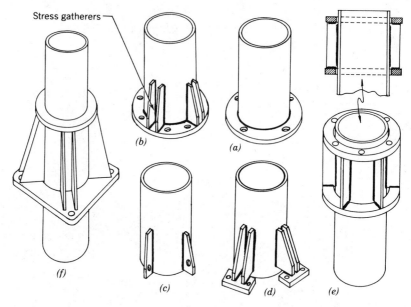

FIGURE 11-4. Pipe flange supports. [From Ref. 2, p. 4.7-5.]

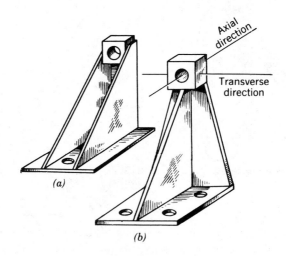

FIGURE 11-5. Welded bracket to hold bearings. [From Ref. 18, p. 328.]

transverse loads than bracket (a) due to the presence of the web that is perpendicular to the bearing axis. Bracket (a) will be stronger in carrying loads parallel to the bearing axis. In each case, force flows from the bearing block down to the base plate. The triangular members gather the force and transmit it down to the base.

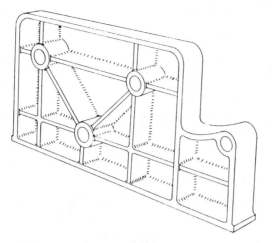

FIGURE 11-6. Cast iron side frame. [From Ref. 2, p. 4.4-1.]

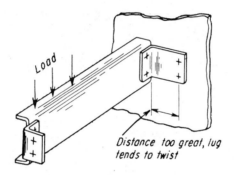

Poor design

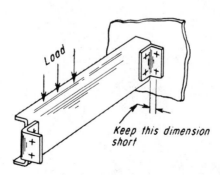

Improved design

FIGURE 11-7. Improved design obtained by reducing mounting lug length. [From Ref. 9, p. 108.]

The tetrahedral-triangle principle could be used to explain and improve the efficiency of these force gathers.

Diagonal bracing is used in the cast-iron side frame pictured in Figure 11-6. This bracing between bearings allows the force to flow directly between the frame-shaft load points. The bracing serves as direct channels for the flow of force.

The three methods given for obtaining high joint efficiency are based on the joint design principles stated in Chapter 10. One of these principles was to keep the bridges as short as possible. Figure 11-7 contains a poor cantilever beam joint design where the "keep the bridges as short as possible" principle has been used to provide an improved design. The shorter bridge reduces the tendency for the lug to twist.

Figure 11-8 pictures a second set of "poor and improved" cantilever beam joint designs. In the improved design, a large number of fasteners are widely and evenly distributed over the joint 1. The widely spaced fasteners produce a larger joint

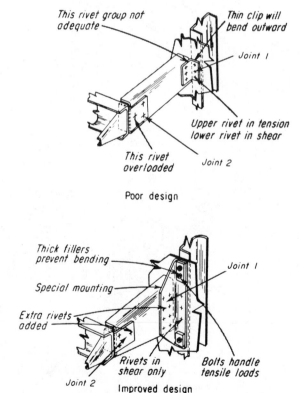

FIGURE 11-8. Improved design obtained by increasing the resistance of the joint to bending. [From Ref. 9, p. 108.]

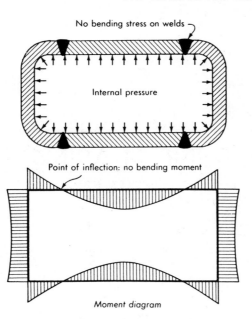

FIGURE 11-9. Welding at points of inflection where no bending moment exists. [From Ref. 2, p. 2.6-7.]

section modulus, and the flow can be transmitted more uniformly from the column to the tapered beam. Force-gathering channels are also employed. Force flows rather uniformly from the column into the beam, and the beam tapers down in size to match the size of joint 2.

The methods mentioned earlier are used where high strength-to-weight joints are needed. There are, however, other schemes that under certain circumstances can be employed in joint design. For example, consider the joint design of Figure 11-9.

The rectangular-shaped vessel is internally pressurized. The bending moment diagram for the pressure loading is shown in the figure. The vertical sides of the vessel see an almost uniform bending moment. But each of the horizontal sides have two points of inflection where no bending moment and hence no bending stress exists. A well-designed vessel of this type could be constructed so as to have the welded joints at the inflection points. This construction is an example of the *principle of joint placement* which states that the joint should be at a position where the member is least severely stressed.

In general, an application of the principle of force flow can initially be used to visualize the stress distribution. Employing force flow is the starting point for a good joint design. Following force flow, principles of strength of materials can be used to determine the location of the inflection points and the areas for joint placement.

12

FURTHER INCREASING BODY AND JOINT EFFICIENCY

Additional principles can be employed to further increase the body and joint efficiency in machine and structural parts. *Supplementary structural parts* can be added to prevent shifts to weaker stress patterns. *Floating parts* can be employed to separate inefficient bending moments from torsional moments. *Elastic matching* can be used to redistribute concentrated flows of force, and *shape refinement* can be employed to improve force flow.

12-1 SUPPLEMENTARY STRUCTURAL SHAPES

Supplementary structural shapes can be used to further increase joint efficiency. Figure 12-1 shows structural stiffeners employed on flat panels to prevent buckling and undue stress in the corner joint of the rectangular bin. Design 1 is better than design 2 since the stiffening capacity of the flange has been used to the maximum advantage in design 1.

Figure 12-2 pictures a support for a tank with a dished bottom. The force from the weight of the tank is transmitted through the support and down into the ground. Any of the shearing ribs could be thought of as supplementary structures that would increase the joint efficiency by preventing an inefficient stress pattern to form in the support. The shearing ribs prevent buckling of the supports by acting as anti-buckling stiffeners, thus preventing the development of inefficient stress patterns.

156

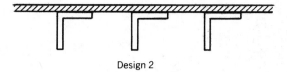

Design 2

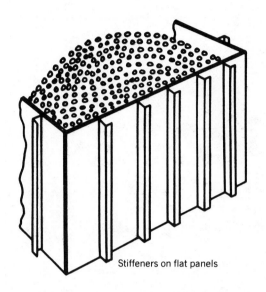

Stiffeners on flat panels

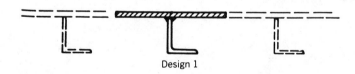

Design 1

FIGURE 12-1. Stiffeners used on flat panels. [From Ref. 2, p. 4.3-7.]

12-2 FLOATING OR SEMIFLOATING PARTS

The floating axle is a good example where a floating or semifloating part is used to increase joint efficiency. In the floating shaft, the shaft is mounted in bearings attached to a housing. The housing carries all bending moments so that the axle only has to transmit motor output torque.

Figure 12-3 shows a floating shaft where torque is transmitted between a motor and a gear-box, but the two flexible couplings limit the transmission of a bending moment.

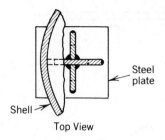

Steel
plate

Shell

Top View

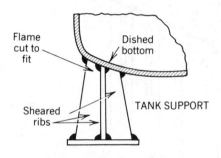

Flame
cut to
fit

Dished
bottom

Sheared
ribs

TANK SUPPORT

FIGURE 12-2. Welded tank support connection. [From Ref. 19, p. 334.]

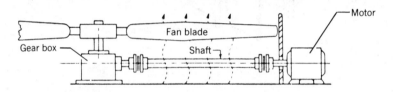

Motor

Fan blade

Gear box

Shaft

FIGURE 12-3. Full-floating shaft and couplings for a cooling tower installation. [From Ref. 20, p. 312.]

12-3 ELASTIC-MATCHING PRINCIPLE

The posterior portion of the human body might be used in an example of elastic matching. When a person sits in a chair, the body rear conforms to the shape of the chair seat. And this portion of the body distributes the body weight to best match the chair contour, hence making it more comfortable to sit.

Figure 12-4 shows a roller supported on each end with a pillow block. The roller is uniformly loaded and bends as exaggerated in the figure. Each roller shaft end rotates through an angle, θ, and this rotation must be matched at the interface by the pillow block bearings. High loads would be induced in the bearings if the bearing supports were stiff and all the rotation had to be matched by the bearings.

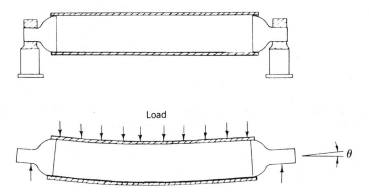

FIGURE 12-4. Roller loaded with a uniform pressure and supported by two fixed bearings. [From Ref. 2, p. 3.5-1.]

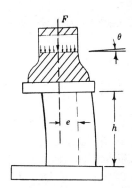

FIGURE 12-5. Bearing support shifted from the center of the bearing by an amount *e*. [From Ref. 2, p. 3.5-1.]

Ideally, the bearing *supports* should be designed to match elastically the rotation of the roller shaft ends. This can be done by fabricating the pillow block support to allow the bearing rotation to conform with shaft rotation. Specifically, this can be effected by offsetting the block a distance *e* from the center of the bearing (see Fig. 12-5). This offset allows bending flexibility in the direction of rotation for the supports and thus reduces the rotational requirements (and therefore stress) that would otherwise be imposed on the bearings (see Fig. 12-6). This means a more uniform stress in the bearing joint element.

Another example where the principle of elastic matching should be employed is revealed in Figure 12-7. Here a bent strap should be used rather than a straight strap which is much stiffer than the bent member. Because the strap is in parallel with the joint, the stiffer strap carries most of the load, inducing large stresses at the strap ends which could cause a crack to develop. The stiffness of the strap should ideally be matched with the stiffness of the bent member, so that force flows through both in proportion to their strength.

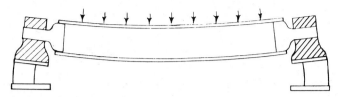

FIGURE 12-6. Improved alignment created by elastic matching. [From Ref. 2, p. 3.5-1.]

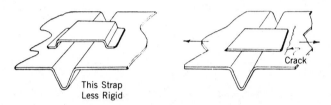

This Strap
Less Rigid

Crack

FIGURE 12-7. Matching stiffness of strap to stiffness of bent plate joint to reduce stress concentration. [From Ref. 2, p. 3.2-9.]

12-4 SHAPE REFINEMENT PRINCIPLE

Notches and Grooves

The shape refinement principle could be called a material removal principle because in many cases higher strength-to-weight ratios can be achieved by removal of material. This principle can be applied to design both body parts and joint elements. For example, consider a severely notched plate. The force-flow principle reveals that the force flow would not anticipate the abrupt notch (see Fig. 12-8). The flow approaching the notch would be forced to deviate drastically from a straight line to avoid the notch, and the more rapid the change in flow line slope, the higher the stress will be.

Changing the flow in the notch area gradually requires a signal upstream that there exists a large stress concentration ahead. This signal can take the form of

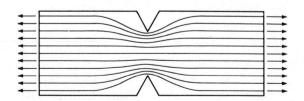

FIGURE 12-8. Flow of force in a notched bar loaded in tension. [From Robert C. Juvinall, *Engineering Considerations of Stress, Strain, and Strength*, McGraw-Hill, New York, 1967, p. 238. Reproduced by permission.]

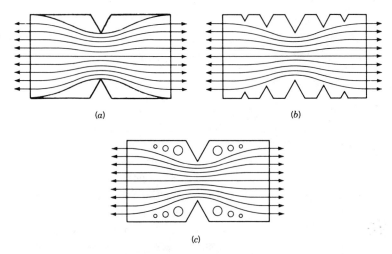

FIGURE 12-9. Removal of excess material to reduce the stress concentration. (*a*) Removal of undesired material, (*b*) cutting notches, (*c*) drilling holes.

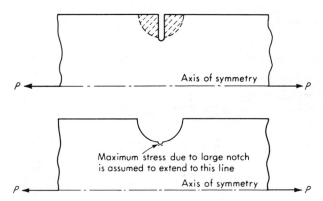

FIGURE 12-10. Single notch more severe that double notch obtained by material removal. [From Robert C. Juvinall, *Engineering Considerations of Stress, Strain, and Strength*, McGraw-Hill, New York, 1967, pp. 257, 258. Reproduced by permission.]

notches or holes that gradually increase in size (see Fig. 12-9). The use of the notches or holes tells the flow that there is a stress concentration downstream and causes the flow lines to change direction sooner and more gradually. A process of removing material by drilling holes or cutting notches that aids in reducing the severity of a stress concentration is an example of the shape refinement principle. Obviously, material could also be added to refine shapes and reduce stresses.

In Figure 12-10, a groove is cut into a member having a load, P, applied at the axis of symmetry. The theoretical stress concentration for this case would be extremely high. By enlarging the groove into a semicircular shape, we find that a much lower stress concentration value would exist, even with a little notch in the bottom. The material removal/shape refinement helps the flow of stress anticipate the change in the cross-sectional area, and the maximum stress in the area of the notch is lowered considerably.

Shoulder Fillets

Figure 12-11 gives shape refinements that could be used for step shafts. Clearly (a) is the worse case. Case (b) is an improvement because of the fillet radius that provides a gradual shaft size transition. Case (c) provides a groove in the large shaft just before the fillet radius. This groove helps channel the stress flow through the step. The groove in (d) reduces the stress concentration as compared with (a) and its location allows the use of the larger shaft as an accurate positioning stop. Cases (e) and (f) use grooves and holes to reduce stress concentration.

Holes

A shaft with a drilled hole is shown in Figure 12-12. Obviously, a stress concentration exists because of the hole. The design engineer would like to know what can be done to reduce the effect of the concentration. One solution would be to

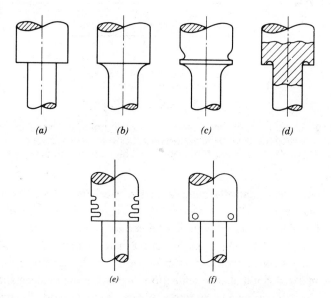

FIGURE 12-11. Removal of material to reduce the stress concentration in a stepped shaft. [From Robert C. Juvinall, *Engineering Considerations of Stress, Strain, and Strength*, McGraw-Hill, New York, 1967, p. 249. Reproduced by permission.]

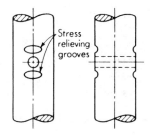

Stress relieving grooves

FIGURE 12-12. Stress concentration at hole reduced by grooves. [From Robert C. Juvinall, *Engineering Considerations of Stress, Strain, and Strength*, McGraw-Hill, New York, 1967, p. 250. Reproduced by permission.]

TAPERED ENDS AND COTTER HOLES

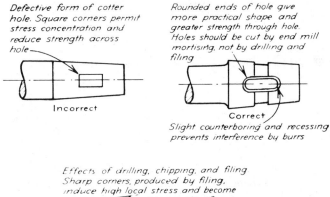

Defective form of cotter hole. Square corners permit stress concentration and reduce strength across hole

Incorrect

Rounded ends of hole give more practical shape and greater strength through hole. Holes should be cut by end mill mortising, not by drilling and filing

Correct

Slight counterboring and recessing prevents interference by burrs

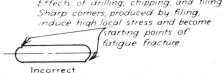

Effects of drilling, chipping, and filing Sharp corners, produced by filing, induce high local stress and become starting points of fatigue fracture

Incorrect

FIGURE 12-13. Cotter hole and shaft hole shape modifications. [From Ref. 10, p. 303.]

mill stress relieving grooves to help the stress flow anticipate the hole. This would be a relatively simple shape refinement that would increase the strength of the part.

Sharp corners in cotter holes have to be avoided (see Fig. 12-13), and stress-relieving grooves near holes in shafts will reduce local stress concentrations. An enlarged diameter around a shaft hole also reduces local stress concentrations and will increase shaft fatigue strength (see Fig. 12-14).

Lightening holes can also be used to increase the strength-to-weight ratio of a part by removing material from unstressed sections (see Fig. 12-15).

Keyseats and Splines

Figure 12-16 shows another example of the effect of shape on the stress concentration factor. Stress concentration factors for the sled runner and profile keyways

HOLES IN SHAFTS

Material around mouth of
hole subjected to excessive
stress Resistance of material
to fatigue failure is
diminished

Stress relieving grooves
cut on each side of hole
reduce local stresses
and increase fatigue
resistance of shaft

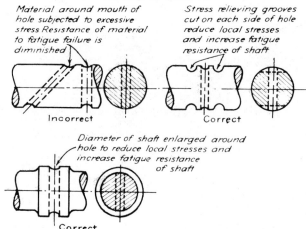

Incorrect

Correct

Diameter of shaft enlarged around
hole to reduce local stresses and
increase fatigue resistance
of shaft

Correct

FIGURE 12-14. Shaft hole shape refinements. [From Ref. 10, p. 303.]

FIGURE 12-15. The use of lightening holes in unstressed webs. [From Ref. 18, p. 328.]

Fatigue-stress-concentration factors K_f; typical values for keyways in solid round steel shafts*

Steel	Profiled keyway		Sled-runner keyway	
	Bending	Torsion	Bending	Torsion
Annealed (less than 200 Bhn)	1.6	1.3	1.3	1.3
Quenched and drawn (over 200 Bhn)	2.0	1.6	1.6	1.6

Note: Nominal stresses should be based on the section modulus for the total shaft section.

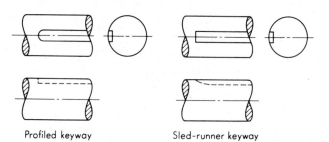

Profiled keyway

Sled-runner keyway

FIGURE 12-16. Effect of keyway shape on fatigue stress concentration. [From Robert C. Juvinall, *Engineering Considerations of Stress, Strain, and Strength*, McGraw-Hill, New York, 1967, p. 252. Reproduced by permission.]

depend on the shaft size, depth of keyway, keyway fillet radius, material used, and load condition. In general, the harder the material is, the higher the stress concentration will be. For loading, bending produces a higher stress concentration than for torsion. But the choice of keyway for bending shows that the profiled keyway shape gives a higher stress concentration than that achieved with the sled-runner keyway shape. Application of the shape refinement principle dictates the choice of sled-runner keyway for bending, whereas for torsion either keyway could be employed.

Shape refinements that can be made in keyways and splines are pictured in Figure 12-17. Stud and shoulder shape refinements are shown in Figure 12-18. In general, a larger diameter and rounded corners are used to obtain lower stresses.

Press-Fitted or Shrink-Fitted Members

Figure 12-19a shows a cylindrical-shaped collar press fit onto a shaft. Because of the press fit, compressive stresses exist in the shaft under the collar. At the edges of the collar there are stress concentrations, whereas in the shaft a small distance from the edges there are smaller press fit stresses. The change from compressive

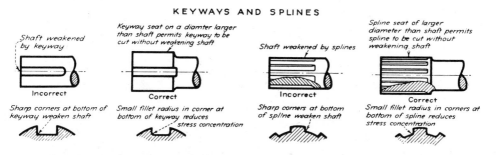

FIGURE 12-17. Keyway and spline shape refinements. [From Ref. 10, p. 303.]

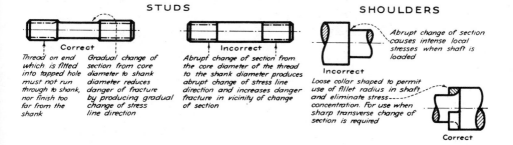

FIGURE 12-18. Stud and shoulder shape refinements. [From Ref. 10, p. 302.]

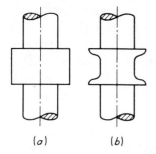

FIGURE 12-19. Removal of collar material to produce lower-stress concentration value. [From Robert C. Juvinall, *Engineering Considerations of Stress, Strain, and Strength*, McGraw-Hill, New York, 1967, p. 250. Reproduced by permission.]

(*a*)   (*b*)

SHRINK AND PRESS FITS

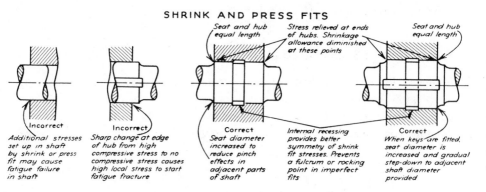

FIGURE 12-20. Press and shrink fit shape refinements. [From Ref. 10, p. 303.]

stress to no stress in the exposed portion of the shaft surface causes stress concentrations in the shaft portion located under the collar edges.

Shape refinement can reduce this stress concentration. Collar material can be removed as shown in Figure 12-19*b* to reduce the stress concentration and prolong shaft life. The exact result of this shape change will depend on shaft and collar sizes, materials, and loading, and environmental conditions.

Press and shrink fit part shapes can be improved by an application of the force-flow principle. By increasing the seat diameter as shown in Figure 12-20, we can reduce the pinching effect in the adjacent parts of the shaft, and to mate the surfaces better we can use an internal recess. This recess will prevent fulcrum rocking for imperfect fits.

Bolt and Nut

The shear stress in a tapered nut is less than that in an untapered nut. Figure 12-21 pictures a regular nut and the stress distribution in the nut threads. Figure 12-22

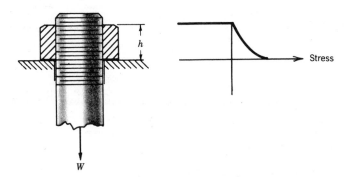

FIGURE 12-21. Nut and bolt and associated stress distribution in the threads of the nut. [From Ref. 12, p. 158.]

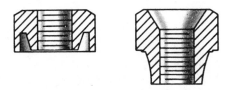

FIGURE 12-22. Shape-refined nuts. [From Ref. 12, p. 158.]

BOLT HEADS

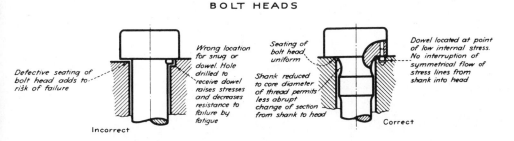

FIGURE 12-23. Bolt head shape improvements. [From Ref. 10, p. 302.]

shows two nut shapes that will produce a more efficient thread stress distribution. The shape refinement follows from an application of the elastic matching and the force-flow principles.

Bolt head shapes can also be improved to produce higher strength-to-weight bolts. Figure 12-23 shows shape improvements that can be made in bolt heads. Additional shape refinements that can be used in nuts are listed in Figure 12-24.

NUTS

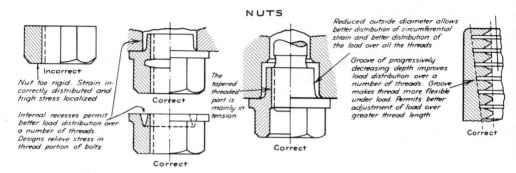

Incorrect

Nut too rigid. Strain incorrectly distributed and high stress localized

Internal recesses permit better load distribution over a number of threads. Designs relieve stress in thread portion of bolts

Correct

Correct

The tapered threaded part is mainly in tension

Correct

Reduced outside diameter allows better distribution of circumferential strain and better distribution of the load over all the threads

Groove of progressively decreasing depth improves load distribution over a number of threads. Groove makes thread more flexible under load. Permits better adjustment of load over greater thread length

Correct

FIGURE 12-24. Nut shape refinements. [From Ref. 10, p. 302.]

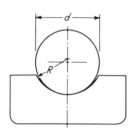

FIGURE 12-25. Ball and ball-bearing race. [From Robert C. Juvinall, *Engineering Considerations of Stress, Strain, and Strength*, McGraw-Hill, New York, 1967, p. 391. Reproduced by permission.]

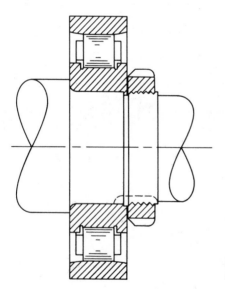

FIGURE 12-26. Cylindrical roller-bearing cross section.

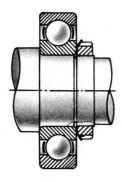

FIGURE 12-27. Ball-bearing cross section. [From Ref. 12, p. 265.]

Ball Bearing

Figure 12-25 pictures a ball on a bearing race. The radius of the race is made as close as possible to the radius of the ball after considering the effect of sliding. This shape refinement follows from an application of the mating surface and the force-flow principles. These principles also suggest the use of a cylindrical roller bearing (Fig. 12-26) rather than a ball bearing (Fig. 12-27) for carrying heavier loads.

13

EFFECT OF
RELATIVE STIFFNESS
ON LOAD DISTRIBUTION

The principle of relative stiffness, which follows from the concept of force flow, describes the way in which the load is shared by an assembly of several machine members or structural elements of different stiffnesses. This principle can be defined in terms of two corollaries.

13-1 PRINCIPLE OF RELATIVE STIFFNESS

Corollary 1: For a stiff member in parallel with a flexible member, the stiff member carries most of the load. Force flux flows in the stiffer member in proportion to the relative stiffnesses.

Consider a rigid horizontal plate as shown in Figure 13-1 which is supported by members 1 and 2 that have spring constants k_1 and k_2. The plate, which is loaded with force P, is guided vertically downward so that it remains horizontal and the deflection of both members is the same. The load carried by member 1 is

$$P_1 = \frac{k_1 P}{(k_1 + k_2)}$$

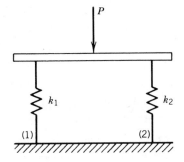

FIGURE 13-1. Members in parallel.

The load carried by member 2 is

$$P_2 = \frac{k_2 P}{(k_1 + k_2)}$$

If member 1 is much stiffer than member 2; that is, $k_1 \gg k_2$, then the load carried by member 1 is much greater than that carried by member 2.

Corollary 2: For a stiff member in series with a flexible member, the stiff member deflects very little relative to the flexible member although it carries the same amount of flux.

Consider a rigid horizontal plate as shown in Figure 13-2 which is supported by members 3 and 4. The plate is loaded with force P and guided downward. The load carried by each member is the same and equals the force P. The deflection of member 3 is

$$\delta_3 = \frac{P}{k_3}$$

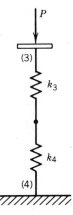

FIGURE 13-2. Members in series.

and the deflection of member 4 is

$$\delta_4 = \frac{P}{k_4}$$

If member 3 is much stiffer than member 4; that is, $k_3 \gg k_4$, then the deflection of member 3 is much less than that of member 4.

In the next sections, we consider several examples that illustrate the principle of relative stiffness and demonstrate how force flux flows in machine components and structures. The first example deals with a type of modular conveyor belting used to transport shrimp for freeze drying. The second example studies a concrete structure loaded at a center column. A third example reveals the load distribution in a roller chain wrapped 180° around a sprocket. The fourth example treats the distribution of torsion load between a splined shaft and a hub. The fifth example shows a spring model of a timing belt engaged with a sprocket. And the final example presents the load distribution for two general types of threaded connectors.

13-2 LOAD DISTRIBUTION IN MODULAR BELTING REINFORCED WITH STEEL LINKS

Figure 13-3a shows a length of modular belting that has its ends reinforced with steel links. One of the steel links having width H and thickness t is also shown. The belting system can be analyzed by the spring model shown in Figure 13-3b. The steel links can each be modeled by a spring of stiffness k_1. The stiffness of the plastic links can be analyzed by a spring of stiffness k_2.

The stiffness k_1 of the steel link is given by

$$k_1 = \left(\frac{AE}{L}\right)_{\text{steel}} = \frac{[(0.378)(0.067)][30 \times 10^6]}{1.5}$$

$$= 5.2 \times 10^5 \text{ lbf/in}$$

The stiffness of the plastic links is given by

$$k_2 = 8\left(\frac{AE}{L}\right)_{\text{plastic}} = \frac{8[(0.378)(0.067)][4 \times 10^5]}{1.5}$$

$$= 5.6 \times 10^4 \text{ lbf/in}$$

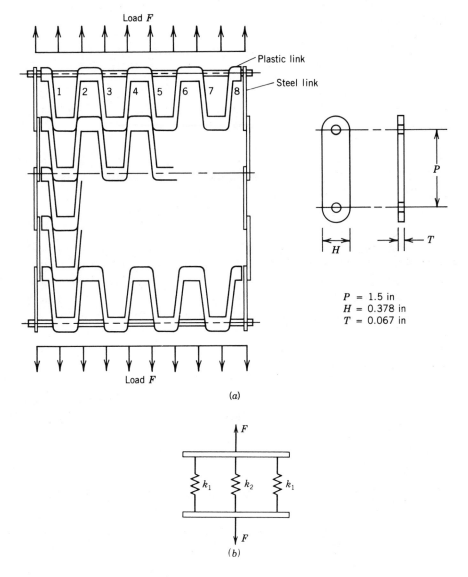

$P = 1.5$ in
$H = 0.378$ in
$T = 0.067$ in

(a)

(b)

FIGURE 13-3. (a) Modular conveyor belting with steel end link, (b) modular belting spring model.

The load carried by the two steel links is

$$P_1 = \left[\frac{2k_1}{(2k_1 + k_2)} \right] F = 0.95F$$

The load carried by the eight plastic links is

$$P_2 = \frac{k_2 F}{(2k_1 + k_2)}$$

The two steel links are stiffer than the plastic links; the steel links carry 95 percent of the total load.

13-3 DISTRIBUTION OF LOAD IN CONCRETE STRUCTURE WITH CENTRALLY LOADED COLUMN

Figure 13-4 shows a concrete structure with a centrally loaded column. All members of this structure are made of concrete and have a 24-in^2 cross section. The load transmitted through members 1 and 4 are equal since these members are in series. Member 2 and the two series combinations of members 1 and 4 and members 3 and 5 are in parallel. The members 1, 2, and 3 are in compression (an efficient stress pattern), whereas members 4 and 5 are in bending and transverse shear (inefficient stress patterns).

The flow of the force P through the structure can be analyzed by using the spring model shown in Figure 13-5a.

The stiffness of members 1, 2, and 3 are

$$k_1 = k_2 = k_3 = \frac{AE}{L} = \frac{[(24)(24)][2 \times 10^6]}{72} = 1.6 \times 10^7 \text{ lbf/in}$$

We assume that members 4 and 5 are under pure bending and neglect their deflection due to shear. The stiffness of members 4 and 5 due to bending (neglecting shear deflections) is

$$k_4 = k_5 = \frac{3EL}{L^3} = \frac{3[2 \times 10^6][27,648]}{72^3} = 4.4 \times 10^5 \text{ lbf/in}$$

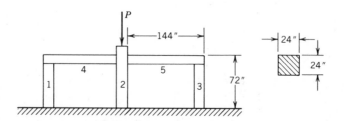

FIGURE 13-4. Concrete structure with centrally loaded column.

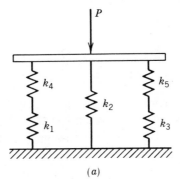

(a)

Given system:

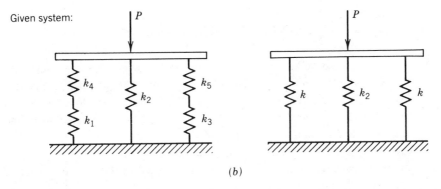

(b)

FIGURE 13-5. (a) Concrete structure spring model, (b) spring model with equivalent member model.

where

$$I = \frac{bh^3}{12} = \frac{[24][24]^3}{12} = 27,648 \text{ in}^4$$

The two series combinations of members 1 and 4 and members 3 and 5 can be replaced by an equivalent member with spring constant k, where

$$k = \frac{k_1 k_4}{(k_1 + k_4)} = \frac{k_3 k_5}{(k_3 + k_5)} = \frac{(1.6 \times 10^7)(4.4 \times 10^5)}{(1.6 \times 10^7 + 4.4 \times 10^5)} = 4.3 \times 10^5 \text{ lbf/in}$$

Figure 13-5b shows the given system and the equivalent member system. The load P_2 carried by member 2 is

$$P_2 = \left[\frac{k_2}{(k_2 + 2k)} \right] P = 0.95P$$

The load carried by the remainder of the structure is $0.05P$.

Since member 2 is under compression, which is an efficient stress pattern, and the other paths of force flow are under bending, which is an inefficient stress pattern, member 2 is stiffer than combined members 1 and 4 and members 3 and 5. As shown in the equation for P_2, member 2 carries 95 percent of the total transmitted load.

13-4 CHAIN AND SPROCKET LOAD DISTRIBUTION

In chain and sprocket power transmission equipment, the chain is in contact with a number of sprocket teeth and each tooth of the sprocket is subjected to a certain percentage of the total transmitted load. The distribution of the tooth load (and also the chain tension) depends on the relative stiffness of the sprocket teeth and the chain links and also on several other parameters of the drive.

The following results that are based on a spring model analysis for the load distribution in chains and in sprocket teeth can be used to illustrate the effect of their relative stiffness. For additional information see Ref. 45.

The analysis presented is for a chain and a double pitch sprocket of the type shown in Figures 13-6 and 13-7. The load in the chain is a maximum and is equal to F on the tight side and zero on the slack side.

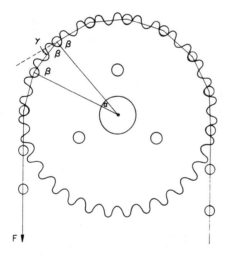

FIGURE 13-6. Force acting on chain wrapped on double pitch fixed sprocket. [From Ref. 45, p. 136.]

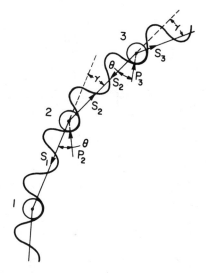

FIGURE 13-7. Free-body diagrams for chain rollers 2 and 3. [From Ref. 45, p. 136.]

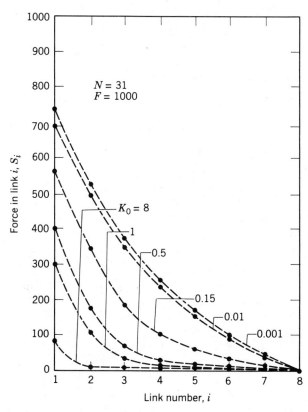

FIGURE 13-8. Tensile-force distribution in chain links predicted for different values of K_0 by spring model analysis. [From Ref. 45, p. 138.]

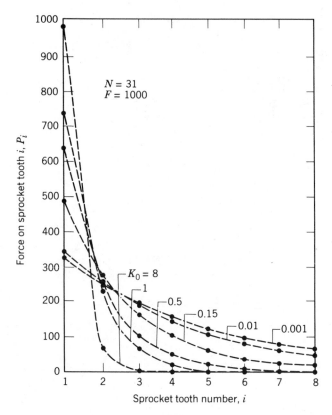

FIGURE 13-9. Forces on double pitch sprocket teeth predicted for different values of K_0 by spring model analysis. [From Ref. 45, p. 138.]

We define K_t as the spring constant of a single sprocket tooth; K_1 as the spring constant of a single chain link; and $K_0 = K_t/K_1$.

For a double pitch sprocket with eight rollers in contact, the following results were derived [45]. As can be seen in Figures 13-8 and 13-9, when the sprocket teeth are very rigid as compared to the chain, the first tooth in contact carries most of the load and the other sprocket teeth are comparatively lightly loaded. With a very rigid chain and flexible sprocket teeth (i.e., $K_0 \cong 0$), the tooth load is relatively uniformly distributed among all sprocket teeth.

13-5 THE DISTRIBUTION OF LOAD BETWEEN A SPLINED SHAFT AND A HUB

Consider the case of a torque that is transmitted between a splined shaft and a hub, both members being elastic. The distribution in the axial direction of the torque load transmitted by the splines depends on the relative stiffness of the two members.

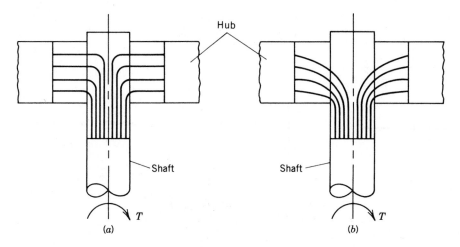

FIGURE 13-10. Splined shaft and hub of different relative stiffnesses. (*a*) rigid shaft, flexible hub, (*b*) flexible shaft, rigid hub.

We assume that the external torque exerted on the hub is uniformly distributed in the circumferential direction; if the shaft is very rigid and the hub flexible, then the torque transmitted by the splines will be uniformly distributed in the axial direction. If the hub is rigid, with a very flexible shaft, then most of the torque will be transmitted through one end of the splines as shown in Figure 13-10.

13-6 LOAD DISTRIBUTION IN TIMING BELTS

The distribution of the load between the timing belt and the toothed wheel shown in Figure 13-11 depends on their relative stiffness. The variation of the tooth load along the arc of contact can be studied by using the spring model as illustrated in Figure 13-12.

We define F_2 as the tension on the tight side, F_1 as the tension on the slack side, and n as the number of teeth in contact.

For the case of a very rigid belt and a flexible toothed wheel, the distribution of the tooth load is uniform and is given by

$$P_i = (F_2 - F_1)/n \qquad (1 \leq i \leq n)$$

At the other extreme, that is, when the toothed wheel is rigid and the belt very elastic, almost the entire load is taken by the first and the last tooth in contact and

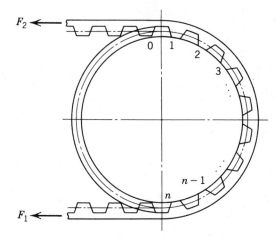

FIGURE 13-11. Timing belt and pulley with belt tooth numbering. [From Ref. 46, p. 208.]

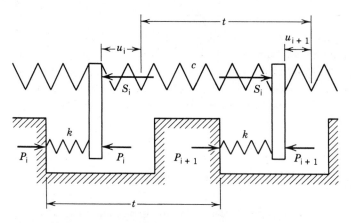

FIGURE 13-12. Spring model of a timing belt. [From Ref. 46, p. 209.]

the other teeth in the arc of contact are very lightly loaded. Ideally, this load distribution is given by the following set of equations:

$$P_1 = F_2, \quad P_n = -F_1, \quad \text{and} \quad P_i = 0, \quad 2 \leq i \leq n - 1$$

13-7 LOAD DISTRIBUTION IN A THREADED CONNECTION

The distribution of the load along the threaded portion of a connector depends on the relative stiffness of the two parts and can be studied by using a spring model analysis as done for the earlier cases.

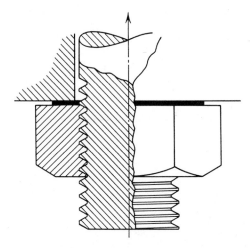

FIGURE 13-13. Compression case—nut and stud or bolt. [From Ref. 47, p. 422.]

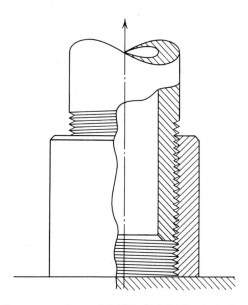

FIGURE 13-14. Tension case—pipe union with straight threads. [From Ref. 47, p. 422.]

Two examples of threaded connections, one in which the bolt is under compression and the other in which it is under tension are illustrated in Figures 13-13 and 13-14.

Figure 13-15 gives the distribution of the load along the threaded portion for a steel bolt and nut under tension, whereas Figures 13-16 and 13-17 give the distribution for the compression case with a steel stud and body and also with a steel stud and an aluminum body, respectively.

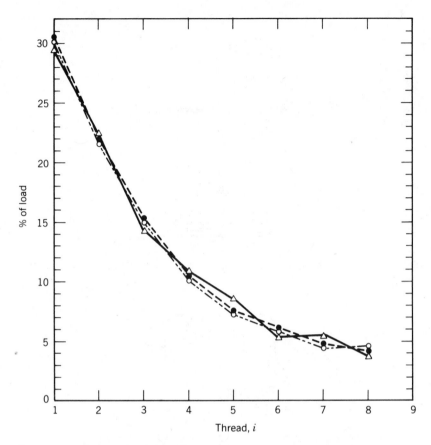

FIGURE 13-15. Case I—comparison of photoelastic, theoretical, and spring model of 1-in diameter steel bolt and nut with 8 Whitworth threads/in. Compression condition with eight engaged threads. Photoelastic model △-△, theoretical ○-○, spring model ●-●. [From Ref. 47, p. 427.]

13-8 LOAD-SHIFT PHENOMENON

One of the idealizations of statics is that bodies are rigid. This means that a body does not deform when loads are applied; that is, any two points of the body remain at a fixed distance from one another. The rigid body idealization is excellent if the deformation is very small or if great accuracy is not required or if the deflection does not cause a significant change in the analytical results. However, deformation can cause a change in the direction and position of applied loads or in the position and direction of reacting forces. Assessing the effect of deformation accurately can be a difficult task. Yet, if more accurate results are needed, more realistic models that account for deformation must be used. One approach that is widely used is to

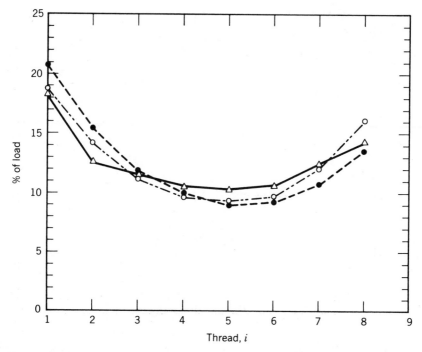

FIGURE 13-16. Case III—comparison of finite element model, theoretical, and spring model of 1-in steel stud and body with eight UNC threads. Tension condition with eight active threads. Finite element model △-△, theoretical ○-○, spring model ●-●. [From Ref. 47, p. 428.]

take the deformed shape as the shape of the rigid body. This approach will yield good results since the reacting supports will now correspond to the shape of the body that is deformed by loading.

There are numerous examples where the deformation of machine and structural parts causes a shift in load and a redistribution of force flow. Figure 13-18 shows a connecting rod end that transmits a force P. The central portion of the rod is in tension. The portion that clamps to the crankshaft is subjected to bearing and bending stresses.

Inspection of the bolted joints reveals separation at the inside of the joint. This takes place because of the joint and member deflections that occur under load, and this movement changes the stress pattern in the bolts from pure tension to combined tension and bending.

Failure to consider this load shift and stress redistribution in the joint area would produce a computed stress which is smaller than the one that actually exists. The load shift must be recognized as a phenomenon that can create higher stresses in certain cases. Consequently, it is necessary to consider the effect of load shift.

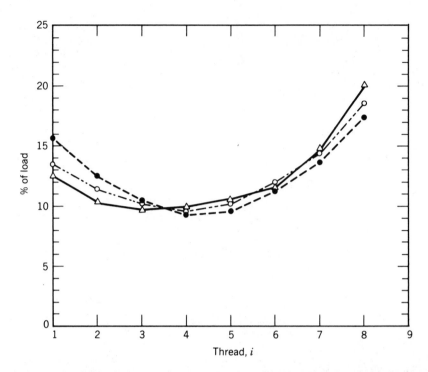

FIGURE 13-17. Case IV—comparison of finite element model, theoretical, and spring model of a 1-in steel stud and aluminum alloy body with eight UNC threads. Tension condition with eight active threads. Finite element model △-△, theoretical ○-○, spring model ●-●. [From Ref. 47, p. 429.]

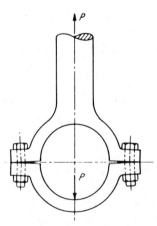

FIGURE 13-18. Exaggerated deflections in axial-loaded connecting rod joint. [From Ref. 4, p. 229.]

Using simple analysis, we can do this best by visualizing the shape of the part under load and then sketching a free-body diagram of the deformed part showing a redistributed loading. The analysis would be performed with the deformed free-body diagram.

APPENDIX **1**

SI PREFIXES

Factor	Prefix	Symbol		Factor	Prefix	Symbol
10^{12}	tera	T		10^{-1}	deci	d
10^{9}	giga[a]	G		10^{-2}	centi	c
10^{6}	mega[a]	M		10^{-3}	milli[a]	m
10^{3}	kilo[a]	k		10^{-6}	micro[a]	μ
10^{2}	hecto	h		10^{-9}	nano	n
10^{1}	deka	da		10^{-12}	pico	p
				10^{-15}	femto	f
				10^{-18}	atto	a

Source: Chester H. Page and Paul Vigoureux, eds., *The International System of Units (SI)*, Superintendent of Documents, U.S. Government Printing Office, Washington, D.C. 20402 (Order by SD Catalog No. C 13.10 : 330/2), National Bureau of Standards Special Publication 330, 1972, p. 12.
[a]Preferred.

APPENDIX 2

SI UNITS AND SYMBOLS

Quantity	Name	Symbol	Expressed in Other Units
Length[a]	meter	m	
Mass[a]	kilogram	kg	
Time[a]	second	s	
Temperature[a,b]	kelvin	K	
Plane angle[c]	radian	rad	
Acceleration	meter per second squared	m/s^2	
Angular acceleration	radian per second squared	rad/s^2	
Angular velocity	radian per second	rad/s	
Area	square meter	m^2	
Density	kilogram per cubic meter	kg/m^3	
Energy	joule	J	$N \cdot m$
Force	newton	N	$m \cdot kg \cdot s^{-2}$
Frequency	hertz	Hz	s^{-1}
Heat, quantity of	joule	J	$N \cdot m$
Moment of force	meter newton	$N \cdot m$	
Power	watt	W	J/s
Pressure	pascal	Pa	N/m^2
Specific heat capacity	joule per kilogram kelvin	$J/(kg \cdot K)$	

Quantity	Name	Symbol	Expressed in Other Units
Length[a]	meter	m	
Mass[a]	kilogram	kg	
Time[a]	second	s	
Temperature[a,b]	kelvin	K	
Plane angle[c]	radian	rad	
Speed	meter per second	m/s	
Thermal conductivity	watt per meter kelvin	W/(m · K)	
Velocity	meter per second	m/s	
Viscosity, dynamic	pascal second	Pa · s	
Volume	cubic meter	m^3	
Work	joule	J	N · m

Source: Chester H. Page and Paul Vigoureux, eds., *The International System of Units (SI)*, Superintendent of Documents, U.S. Government Printing Office, Washington, D.C. 20402 (Order by SD Catalog No. C 13.10:330/2), National Bureau of Standards Special Publications 330, 1972, p. 12.
[a]SI base unit.
[b]Celsius temperature is expressed in degrees Celsius (symbol °C).
[c]Supplementary unit.

APPENDIX **3**

CONVERSION FACTOR EQUALITIES LISTED BY PHYSICAL QUANTITY

ACCELERATION
*1 foot/second^2 = 3.048 × 10^{-1} meter/second^2
*1 free fall, standard = 9.806 65 meter/second^2
*1 inch/second^2 = 2.54 × 10^{-2} meter/second^2

AREA
*1 acre = 4.046 856 422 4 × 10^3 meter^2
*1 foot^2 = 9.290 304 × 10^{-2} meter^2
*1 hectare = 1.00 × 10^4 meter^2
*1 inch^2 = 6.4516 × 10^{-4} meter^2
*1 mile^2 (U.S. statute) = 2.589 988 110 336 × 10^6 meter^2
*1 yard^2 = 8.361 273 6 × 10^{-1} meter^2

DENSITY
*1 gram/centimeter^3 = 1.00 × 10^3 kilogram/meter^3
 1 lbm/inch^3 = 2.767 9905 × 10^4 kilogram/meter^3
 1 lbm/foot^3 = 1.601 846 3 × 10^1 kilogram/meter^3
 1 slug/foot^3 = 5.153 79 × 10^2 kilogram/meter^3

ENERGY
 1 British thermal unit (mean) = 1.055 87 × 10^3 joules
*1 erg = 1.00 × 10^{-7} joule

1 foot lbf = 1.355 817 9 joule
*1 kilowatt hour = 3.60 × 10^6 joule
1 ton (nuclear equivalent of TNT) = 4.20 × 10^9 joules
*1 watt hour = 3.60 × 10^3 joule

FORCE
*1 dyne = 1.00 × 10^{-5} newton
*1 kilogram force (kgf) = 9.806 65 newton
*1 kilopound force = 9.806 65 newton
*1 kip = 4.448 221 615 260 5 × 10^3 newton
*1 lbf (pound force, avoirdupois) = 4.448 221 615 260 5 newton
1 ounce force (avoirdupois) = 2.780 138 5 × 10^{-1} newton
*1 pound force, lbf (avoirdupois) = 4.448 221 615 260 5 newton
*1 poundal = 1.382 549 543 76 × 10^{-1} newton

LENGTH
*1 angstrom = 1.00 × 10^{-10} meter
*1 cubit = 4.572 × 10^{-1} meter
*1 fathom = 1.8288 meter
*1 foot = 3.048 × 10^{-1} meter
*1 inch = 2.54 × 10^{-2} meter
*1 league (international nautical) = 5.556 × 10^3 meter
1 light year = 9.460 55 × 10^{15} meter
*1 meter = 1.650 763 73 × 10^6 wavelength Kr 86
*1 micron = 1.00 × 10^{-6} meter
*1 mil = 2.54 × 10^{-5} meter
*1 mile (U.S. statute) = 1.609 344 × 10^3 meter
*1 nautical mile (U.S.) = 1.852 × 10^3 meter
*1 yard = 9.144 × 10^{-1} meter

MASS
*1 carat (metric) = 2.00 × 10^{-4} kilogram
*1 grain = 6.479 891 × 10^{-5} kilogram
*1 lbm (pound mass, avoirdupois) = 4.535 923 7 × 10^{-1} kilogram
*1 ounce mass (avoirdupois) = 2.834 952 312 5 × 10^{-2} kilogram
1 slug = 1.459 390 29 × 10^1 kilogram
*1 ton (long) = 1.016 046 908 8 × 10^3 kilogram
*1 ton (metric) = 1.00 × 10^3 kilogram
1 ton (short, 2000 pound) = 9.071 847 4 × 10^2 kilogram

POWER

Btu (thermochemical)/second = $1.054\,350\,264\,488 \times 10^{+3}$ watt
*1 calorie (thermochemical)/second = 4.184 watt
1 foot lbf/minute = $2.259\,696\,6 \times 10^{-2}$ watt
1 foot lbf/second = 1.355 817 9 watt
1 horsepower (550 foot lbf/second) = $7.456\,998\,7 \times 10^2$ watt
*1 horsepower (electric) = 7.46×10^2 watt

PRESSURE

*1 atmosphere = $1.013\,25 \times 10^5$ newton/meter^2
*1 bar = 1.00×10^5 newton/meter^2
1 centimeter of mercury (0°C) = $1.333\,22 \times 10^3$ newton/meter^2
1 centimeter of water (4°C) = $9.806\,38 \times 10^1$ newton/meter^2
*1 dyne/centimeter^2 = 1.00×10^{-1} newton/meter^2
1 inch of mercury (60°F) = $3.376\,85 \times 10^3$ newton/meter^2
1 inch of water (60°F) = 2.4884×10^2 newton/meter^2
*1 kgf/meter^2 = 9.806 65 newton/meter^2
1 lbf/foot^2 = $4.788\,025\,8 \times 10^1$ newton/meter^2
1 lbf/inch^2 (psi) = $6.894\,757\,2 \times 10^3$ newton/meter^2
*1 millibar = 1.00×10^2 newton/meter^2
1 millimeter of mercury (0°C) = $1.333\,224 \times 10^2$ newton/meter^2
*1 pascal = 1.00 newton/meter^2
1 psi (lbf/inch^2) = $6.894\,757\,2 \times 10^3$ newton/meter^2
1 torr (0°C) = $1.333\,22 \times 10^2$ newton/meter^2

SPEED

*1 foot/minute = 5.08×10^{-3} meter/second
*1 foot/second = 3.048×10^{-1} meter/second
*1 inch/second = 2.54×10^{-2} meter/second
1 kilogram/hour = $2.777\,777\,8 \times 10^{-1}$ meter/second
1 knot (international) = $5.144\,444\,444 \times 10^{-1}$ meter/second
*1 mile/hour (U.S. statute) = 4.4704×10^{-1} meter/second

TEMPERATURE

Celsius = kelvin − 273.15
Fahrenheit = (9/5)kelvin − 459.67
Fahrenheit = (9/5)Celsius + 32
Rankine = (9/5)kelvin

TIME

*1 day (mean solar) = 8.64×10^4 second (mean solar)
*1 hour (mean solar) = 3.60×10^3 second (mean solar)

*1 minute (mean solar) = 6.00×10^1 second (mean solar)
*1 month (mean calendar) = 2.628×10^6 second (mean solar)
*1 year (calendar) = 3.1536×10^7 second (mean solar)

VISCOSITY
*1 centistoke = 1.00×10^{-6} meter^2/second
*1 stoke = 1.00×10^{-4} meter^2/second
*1 foot^2/second = $9.290\ 304 \times 10^{-2}$ meter^2/second
*1 centipoise = 1.00×10^{-3} newton second/meter^2
 1 lbm/foot second = $1.488\ 163\ 9$ newton second/meter^2
 1 lbf second/foot^2 = $4.788\ 025\ 8 \times 10^1$ newton second/meter^2
*1 poise = 1.00×10^{-1} newton second/meter^2
 1 slug/foot second = $4.788\ 025\ 8 \times 10^1$ newton second/meter^2

VOLUME
 1 barrel (petroleum, 42 gallons) = $1.589\ 873 \times 10^{-1}$ meter^3
*1 board foot ($1' \times 1' \times 1''$) = $2.359\ 737\ 216 \times 10^{-3}$ meter^3
*1 bushel (U.S.) = $3.523\ 907\ 016\ 688 \times 10^{-2}$ meter^3
 1 cord = $3.624\ 556\ 3$ meter^3
*1 cup = $2.365\ 882\ 365 \times 10^{-4}$ meter^3
*1 fluid ounce (U.S.) = $2.957\ 352\ 956\ 25 \times 10^{-5}$ meter^3
*1 foot^3 = $2.831\ 684\ 659\ 2 \times 10^{-2}$ meter^3
*1 gallon (U.S. dry) = $4.404\ 883\ 770\ 86 \times 10^{-3}$ meter^3
*1 gallon (U.S. liquid) = $3.785\ 411\ 784 \times 10^{-3}$ meter^3
*1 inch^3 = $1.638\ 706\ 4 \times 10^{-5}$ meter^3
*1 liter = 1.00×10^{-3} meter^3
*1 ounce (U.S. fluid) = $2.957\ 352\ 956\ 25 \times 10^{-5}$ meter^3
*1 peck (U.S.) = $8.809\ 767\ 541\ 72 \times 10^{-3}$ meter^3
*1 pint (U.S. dry) = $5.506\ 104\ 713\ 575 \times 10^{-4}$ meter^3
*1 pint (U.S. liquid) = $4.731\ 764\ 73 \times 10^{-4}$ meter^3
*1 quart (U.S. dry) = $1.101\ 220\ 942\ 715 \times 10^{-3}$ meter^3
 1 quart (U.S. liquid) = $9.463\ 529\ 5 \times 10^{-4}$ meter^3
*1 stere = 1.00 meter^3
*1 tablespoon = $1.478\ 676\ 478\ 125 \times 10^{-5}$ meter^3
*1 teaspoon = $4.928\ 921\ 593\ 75 \times 10^{-6}$ meter^3
*1 ton (register) = $2.831\ 684\ 659\ 2$ meter^3
*1 yard^3 = $7.645\ 548\ 579\ 84 \times 10^{-1}$ meter^3

Source: E. A. Mechtly, *The International System of Units, Physical Constants and Conversion Factors,* NASA SP-7012, Scientific and Technical Information Office, National Aeronautics and Space Administration, Washington, D.C., 1973.
*An exact definition.

CONVERSION FACTOR EQUATIONS

*1 acre = 4.046 856 422 4 × 10^3 meter^2

*1 atmosphere = 1.013 25 × 10^5 newton/meter^2

*1 bar = 1.00 × 10^5 newton/meter^2

1 barrel (petroleum, 42 gallons) = 1.589 873 × 10^{-1} meter^3

*1 board foot (1' × 1' × 1") = 2.359 737 216 × 10^{-3} meter^3

1 British thermal unit (mean) = 1.055 87 × 10^3 joule

*1 bushel (U.S.) = 3.523 907 016 688 × 10^{-2} meter^3

1 calorie (mean) = 4.190 02 joule

*1 carat (metric) = 2.00 × 10^{-4} kilogram

Celsius temperature = kelvin temperature − 273.15

1 degree Celsius = 1 degree kelvin

1 centimeter of mercury (0°C) = 1.333 22 × 10^{+3} newton/meter^2

1 centimeter of water (4°C) = 9.806 38 × 10^1 newton/meter^2

*1 cubit = 4.572 × 10^{-1} meter

*1 day (mean solar) = 8.64 × 10^4 second (mean solar)

1 degree (angle) = 1.745 329 251 9943 × 10^{-2} radian

*1 denier (international) = 1.00 × 10^{-7} kilogram/meter

*1 dyne = 1.00 × 10^{-5} newton

1 erg = 1.00 × 10^{-7} joule

Fahrenheit temperature = $\left(\dfrac{9}{5}\right)$ kelvin temperature − 459.67

1 degree Fahrenheit $= \left(\dfrac{9}{5}\right)$ degree kelvin

Fahrenheit temperature $= \left(\dfrac{9}{5}\right)$ Celsius temperature $+ 32$

1 degree Fahrenheit $= \left(\dfrac{9}{5}\right)$ degree Celsius

*1 fathom $= 1.828\ 8$ meter
*1 fluid ounce (U.S.) $= 2.957\ 352\ 956\ 25 \times 10^{-5}$ meter^3
*1 foot $= 3.048 \times 10^{-1}$ meter
 1 footcandle $= 1.076\ 391\ 0 \times 10^{1}$ lumen/meter^2
*1 free fall, standard $= 9.806\ 65$ meter/second^2
*1 furlong $= 2.011\ 68 \times 10^{2}$ meter
*1 gallon (U.S. dry) $= 4.404\ 883\ 770\ 86 \times 10^{-3}$ meter^3
*1 gallon (U.S. liquid) $= 3.785\ 411\ 784 \times 10^{-3}$ meter^3
*1 grain $= 6.479\ 891 \times 10^{-5}$ kilogram
*1 gram $= 1.00 \times 10^{-3}$ kilogram
*1 hectare $= 1.00 \times 10^{4}$ meter^2
 1 horsepower (550 foot lbf/second) $= 7.456\ 998\ 7 \times 10^{2}$ watt
*1 horsepower (electric) $= 7.46 \times 10^{2}$ watt
*1 hour (mean solar) $= 3.60 \times 10^{3}$ second (mean solar)
*1 inch $= 2.54 \times 10^{-2}$ meter
 1 inch of mercury (60°F) $= 3.376\ 85 \times 10^{3}$ newton/meter^2
 1 inch of water (60°F) $= 2.4884 \times 10^{2}$ newton/meter^2
*1 kilogram mass $= 1.00$ kilogram
*1 kilogram force (kgf) $= 9.806\ 65$ newton
*1 kilopound force $= 9.806\ 65$ newton
*1 kip $= 4.448\ 221\ 615\ 260\ 5 \times 10^{3}$ newton
 1 knot (international) $= 5.144\ 444\ 444 \times 10^{-1}$ meter/second
*1 lbf (pound force, avoirdupois) $= 4.448\ 221\ 615\ 260\ 5$ newton
*1 lbm (pound mass, avoirdupois) $= 4.535\ 923\ 7 \times 10^{-1}$ kilogram
*1 league (international nautical) $= 5.556 \times 10^{+3}$ meter
 1 light year $= 9.460\ 55 \times 10^{15}$ meter
 1 liter $= 1.00 \times 10^{-3}$ meter^3
*1 meter $= 1.650\ 763\ 73$ wavelengths Kr 86
*1 micron $= 1.00 \times 10^{-6}$ meter
 1 mil $= 2.54 \times 10^{-5}$ meter
*1 mile (U.S. statute) $= 1.609\ 344 \times 10^{3}$ meter
*1 millibar $= 1.00 \times 10^{2}$ newton/meter^2
 1 millimeter of mercury (0°C) $= 1.333\ 224 \times 10^{2}$ newton/meter^2
 1 minute (angle) $= 2.908\ 882\ 086\ 66 \times 10^{-4}$ radian

*1 minute (mean solar) $= 6.00 \times 10^{+1}$ second (mean solar)
*1 month (mean calendar) $= 2.628 \times 10^{6}$ second (mean solar)
*1 nautical mile (U.S.) $= 1.852 \times 10^{3}$ meter
 1 ounce force (avoirdupois) $= 2.780\ 138\ 5 \times 10^{-1}$ newton
 1 ounce mass (avoirdupois) $= 2.834\ 952\ 312\ 5 \times 10^{-2}$ kilogram
*1 ounce mass (troy or apothecary) $= 3.110\ 347\ 68 \times 10^{-2}$ kilogram
*1 ounce (U.S. fluid) $= 2.957\ 352\ 956\ 25 \times 10^{-5}$ meter^{3}
*1 pascal $= 1.00$ newton/meter^{2}
*1 pint (U.S. dry) $= 5.506\ 104\ 713\ 575 \times 10^{-4}$ meter^{3}
*1 pint (U.S. liquid) $= 4.731\ 764\ 73 \times 10^{-4}$ meter^{3}
*1 poise $= 1.00 \times 10^{-1}$ newton second/meter^{2}
 1 pound force (lbf avoirdupois) $= 4.448\ 221\ 615\ 260\ 5$ newton
*1 pound mass (lbm avoirdupois) $= 4.535\ 923\ 7 \times 10^{-1}$ kilogram
*1 poundal $= 1.382\ 549\ 543\ 76 \times 10^{-1}$ newton
 1 quart (U.S. dry) $= 1.101\ 220\ 942\ 715 \times 10^{-3}$ meter^{3}
 1 quart (U.S. liquid) $= 9.463\ 592\ 5 \times 10^{-4}$ meter^{3}
 Rankine temperature $= (9/5)$ kelvin temperature
 1 slug $= 1.459\ 390\ 29 \times 10^{1}$ kilogram
*1 stere $= 1.00$ meter^{3}
*1 stoke $= 1.00 \times 10^{-4}$ meter^{2}/second
*1 tablespoon $= 1.478\ 676\ 478\ 125 \times 10^{-5}$ meter^{3}
*1 teaspoon $= 4.928\ 921\ 593\ 75 \times 10^{-6}$ meter^{3}
*1 ton (long) $= 1.016\ 046\ 908\ 8 \times 10^{3}$ kilogram
*1 ton (metric) $= 1.00 \times 10^{3}$ kilogram
 1 ton (nuclear equivalent of TNT) $= 4.20 \times 10^{9}$ joule
*1 ton (short, 2000 pound) $= 9.071\ 847\ 4 \times 10^{2}$ kilogram
 1 torr (0°C) $= 1.333\ 22 \times 10^{2}$ newton/meter^{2}
*1 yard $= 9.144 \times 10^{-1}$ meter
*1 year (calendar) $= 3.1536 \times 10^{7}$ second (mean solar)
*1 year (calendar) $= 3.1536 \times 10^{7}$ second (mean solar)

Source: E. A. Mechtly, *The International System of Units, Physical Constants and Conversion Factors*, NASA SP-7012, Scientific and Technical Information Office, National Aeronautics and Space Administration, Washington, D.C., 1973.

*An exact definition.

PHYSICAL CONSTANTS OF MATERIALS*a,b*

PHYSICAL CONSTANTS OF MATERIALS[a,b]

Material	Approx. Melting Temp., °F	Modulus of Elasticity, E Mpsi	Modulus of Rigidity, G MPsi	Poisson's ratio, ν	Unit wt., ω (lb/in³)	Coefficient of Therm. Expan., α $10^{-6}/°F$	Thermal Conductivity Btu/hr-ft-°F	Specific Heat Btu/lb-°F
Acetal	200[c]	0.4	—	—	0.051	0.5	0.13	0.35
Aluminum alloy	1000	10.3	3.9	0.32	0.10	12.	100	0.22
Brass	—	15.9	6.0	0.33	0.31	11.4	46	0.10
Bronze	—	13.8	5.1	0.35	0.30	10.0	43	0.10
Concrete	—	4.	—	—	0.09	6.0	—	—
Copper	1980	15.7	5.8	0.35	0.32	9.4	205	0.10
Glass	—	50	—	—	0.09	5.0	.5	—
Gold	1950	11.0	3.7	0.42	0.70	7.9	170	0.03
Iron, gray cast	—	15	6.0	0.26	0.26	6.4	26	0.13
Lead	620	2.5	0.8	0.43	0.41	—	21	0.03
Limestone	—	—	—	—	0.09	—	—	—

Material							
Magnesium alloy	1100	6.5	2.4	0.35	0.065	14.5	55	0.28
Nickel alloy	2400	30	11.5	0.30	0.30	7.0	12	0.12
Nylon 6/12	250[c]	0.3	—	—	0.038	0.13	—	—
Nylon 6/12 (30% glass)	300[c]	1.2	—	—	0.047	0.2	0.14	—
PTFE	500[c]	0.2	—	—	0.080	0.5	0.11	0.25
Polycarbonate	275[c]	0.3	—	—	0.043	0.4	0.19	—
Polyethylene	200[c]	0.2	—	—	0.034	0.7	—	0.55
Silver	1760	11.5	4.3	0.37	0.38	10.9	240	0.06
Steel, carbon	2750	30	11.5	0.30	0.28	6.8	27	0.11
Titanium alloy	3100	16.5	6.2	0.33	0.16	4.9	7	0.12
Zinc alloy	750	12	4.5	0.33	0.24	15.0	64	0.11

[a] Values given are representative. Exact values may vary sometimes significantly with composition and processing.

[b] Conversion factors: 1 Mpsi = 6.9 GPa; 1 lb/in^3 = 27.7 Mg/m^3; 1 10^{-6}/°F = 1.8 10^{-6}/°C; 1 Btu/hr-ft-°F = 1.73 W/m-°C; 1 Btu/lb-°F = 4200 J/kg-°C.

[c] Maximum recommended service temperature, °F.

PROPERTIES OF SECTIONS

PROPERTIES OF SECTIONS

Form of Section	Area and Distances from Centroid to Extremities	Moments of Products of Inertia About Central Axes	Radii of Gyration About Central Axes
1. Rectangle	$A = bd$ $y_1 = \dfrac{d}{2}$ $y_2 = \dfrac{b}{2}$	$I_1 = \dfrac{bd^3}{12}$ $I_2 = \dfrac{db^3}{12}$ $I_1 > I_2$ if $d > b$	$r_1 = 0.2887d$ $r_2 = 0.2887b$
2. Triangle	$A = \dfrac{bd}{2}$ $y_n = \dfrac{d}{2}$ $y_m = \tfrac{1}{2}(b + a)$	$I_n = \dfrac{bd^3}{36}$ $I_m = \dfrac{bd(b^2 - ab + a^2)}{36}$ $\theta_{1,2} = \tfrac{1}{2}\tan^{-1}$ $= \dfrac{d(b - 2a)}{(b^2 - ab + a^2 - d^2)}$	$r_n = 0.2357d$ $r_m = 0.2357 \cdot (b^2 - ab + a^2)^{1/2}$

PROPERTIES OF SECTIONS. (Continued).

Form of Section	Area and Distances from Centroid to Extremities	Moments of Products of Inertia About Central Axes	Radii of Gyration About Central Axes
3. Circle	$A = \pi R^2$ $y_1 = R$	$I_1 = \dfrac{\pi R^4}{4}$	$r_1 = \dfrac{R}{2}$
4. Ellipse	$A = \pi ab$ $y_1 = a$ $y_2 = b$	$I_1 = \left(\dfrac{\pi}{4}\right) ba^3$ $I_2 = \left(\dfrac{\pi}{4}\right) ab^3$	$r_1 = \dfrac{a}{2}$ $r_2 = \dfrac{b}{2}$
5. Sector	$A = \alpha R^2$ $y_{1a} = R \cdot \left(1 - \left(\dfrac{2\sin\alpha}{3\alpha}\right)\right)$ $y_{1b} = \dfrac{2R\sin\alpha}{3\alpha}$ $y_2 = R\sin\alpha$	$I_1 = R^4(\alpha + \sin\alpha\cos\alpha$ $- (16\sin^2\alpha/9\alpha))/4$ $I_2 = \dfrac{R^4(\alpha - \sin\alpha\cos\alpha)}{14}$	$r_1 = R\big(1 + (\sin\alpha\cos\alpha/\alpha)$ $- (16\sin^2\alpha/(9\alpha^2))\big)^{1/2}/2$ $r_2 = \dfrac{R(1 - (\sin\alpha\cos\alpha/\alpha))^{1/2}}{2}$

Source: Raymond J. Roark and Warren C. Young, *Formulas for Stress and Strain* (New York: McGraw-Hill Book Company, 1975), p. 64.

APPENDIX **7**

BENDING DIAGRAMS AND FORMULAS

R = reaction	P = concentrated load
M = bending moment	E = modulus of elasticity
W = total distributed load	I = rectangular moment of inertia
w = distributed load per longitudinal unit	L = distance between support
V = total vertical shear	y = deflection

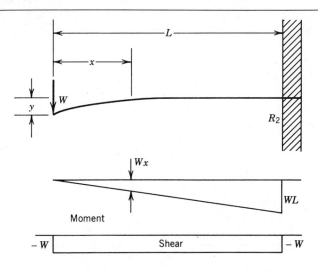

$$R_2 = W$$

$$M_x = -Wx$$

$$M_{max} = -WL \quad (x = L)$$

$$V_x = -W$$

$$y = \left(\frac{WL^3}{3EI}\right)(max)$$

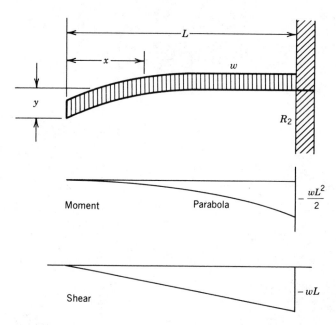

Moment   Parabola $-\dfrac{wL^2}{2}$

Shear $-wL$

$$R_2 = W = wL$$

$$M_x = -\left(\frac{wx^2}{2}\right)$$

$$M_{max} = -\left(\frac{wL^2}{2}\right) \quad (x = L)$$

$$V_x = -wx$$

$$V_{max} = -wL \quad (x = L)$$

$$y = \left(\frac{WL^3}{8EI}\right)(max)$$

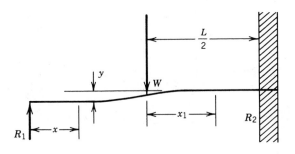

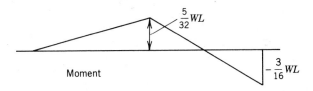

Moment

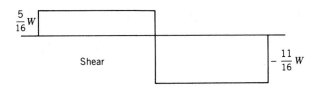

Shear

$$R = \left(\tfrac{5}{16}\right)W, \qquad R_2 = \left(\tfrac{11}{16}\right)W$$

$$M_x = \left(\tfrac{5}{16}\right)Wx$$

$$M_{x1} = WL\left[\left(\frac{5}{32}\right) - \left(\frac{11}{16}\right)\left(\frac{x_1}{L}\right)\right]$$

$$M_{max} = -\left(\frac{3}{16}\right)WL \qquad \left(x_1 = \frac{L}{2}\right)$$

$$V_x = +\left(\tfrac{5}{16}\right)W, \qquad V_{x1} = -\left(\tfrac{11}{16}\right)W$$

$$V_{max} = -\left(\frac{11}{16}\right)W \qquad \left(x = \frac{L}{2} \text{ to } x = L\right)$$

$$y = \left(\frac{7WL^3}{768EI}\right)(\text{max})$$

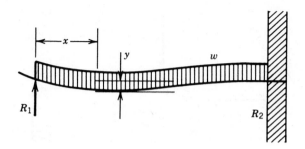

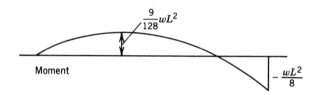

$$R_1 = \left(\tfrac{3}{8}\right)W = \left(\tfrac{3}{8}\right)wL$$

$$R_2 = \left(\tfrac{5}{8}\right)W = \left(\tfrac{5}{8}\right)wL$$

$$M_x = \left(\frac{wx}{2}\right)\left[\left(\frac{3L}{4}\right) - x\right]$$

$$M_{\text{max}} = \left(\frac{9}{128}\right)wL^2 \quad \left(x = \frac{3L}{8}\right)$$

$$M_{\text{max}} = -\left(\frac{wL^2}{8}\right) \quad (x = L)$$

$$V_x = \left[\left(\tfrac{3}{8}\right)wL\right] - wx$$

$$V_{\text{max}} = -\left(\tfrac{5}{8}\right)wL$$

$$y = \left(\frac{WL^3}{185EI}\right)(\text{max})$$

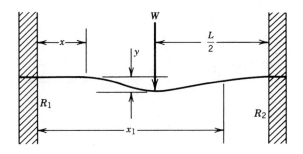

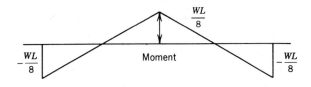

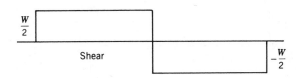

$$R_1 = \left(\frac{W}{2}\right), \quad R_2 = \left(\frac{W}{2}\right)$$

$$M_x = \left(\frac{WL}{2}\right)\left[\left(\frac{x}{L}\right) - \left(\frac{1}{4}\right)\right]$$

$$M_{x1} = -\left(\frac{WL}{2}\right)\left[\left(\frac{x}{L}\right) - \left(\frac{3}{4}\right)\right]$$

$$M_{\text{max}} = \left(\frac{WL}{8}\right) \quad \left(x = \frac{L}{2}\right)$$

$$V_x = \left(\frac{W}{2}\right), \quad V_{x1} = -\left(\frac{W}{2}\right)$$

$$y = \left(\frac{WL^3}{192EI}\right)(\text{max})$$

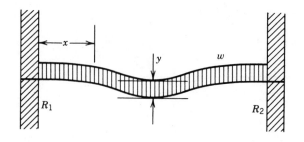

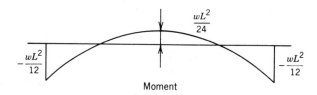

Moment

Shear

$$R_1 = \left(\frac{W}{2}\right) = \left(\frac{wL}{2}\right), \quad R_2 = \left(\frac{W}{2}\right) = \left(\frac{wL}{2}\right)$$

$$M_x = -\left(\frac{wL^2}{2}\right)\left[\left(\frac{1}{6}\right) - \left(\frac{x}{L}\right) + \left(\frac{x^2}{L^2}\right)\right]$$

$$M_{\text{max}} = -\left(\tfrac{1}{12}\right)wL^2 \quad (x = 0 \text{ or } x = L)$$

$$V_x = \left(\frac{wL}{2}\right) - wx$$

$$V_{\text{max}} = \pm\left(\frac{wL}{2}\right)$$

$$y = \left(\frac{WL^3}{384EI}\right)(\text{max})$$

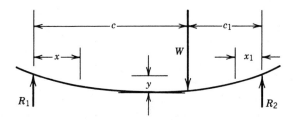

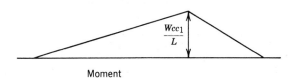

Moment

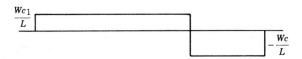

$$R_1 = \left(\frac{Wc_1}{L}\right), \quad R_2 = Wc/L$$

$$M_x = \left(\frac{Wc_1 x}{L}\right), \quad M_{x_1} = (Wcx_1/L)$$

$$M_{\text{max}} = \left(\frac{Wcc_1}{L}\right) \quad (x_1 = c_1 \text{ or } x = c)$$

$$V_x = \left(\frac{Wc_1}{L}\right), \quad V_{x_1} = (Wc/L)$$

$$y = \left(\frac{Wc_1}{3EIL}\right)\left[\left(\frac{c(L + c_1)}{3}\right)\right]^{3/2} (\text{max})$$

$$\text{Max } y \text{ occurs at } x = \left[\frac{c(L + c_1)}{3}\right]^{1/2}$$

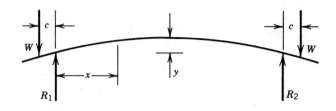

Moment

Shear

$$R_1 = W$$

$$R_2 = W$$

$$M_x = -Wc = \text{const}$$

$$V(W \text{ to } R_1) = -W$$

$$V(R_1 \text{ to } R_2) = 0$$

$$V(R_2 \text{ to } W) = +W$$

$$y_1 = \left(\frac{WcL^2}{8EI}\right)(\text{max})$$

$$y_2 = \left(\frac{Wc^2}{3EI}\right)\left[c + \left(\frac{3L}{2}\right)\right](\text{max})$$

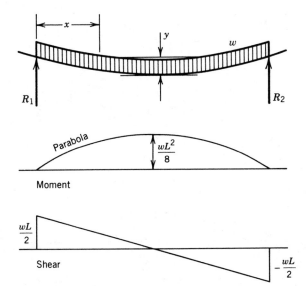

$$R_1 = \frac{W}{2} = \frac{wL}{2}$$

$$R_2 = \frac{W}{2} = \frac{wL}{2}$$

$$M_x = \left(\frac{wx}{2}\right)(L - x)$$

$$M_{max} = \left(\frac{wL^2}{8}\right) \quad \left(x = \frac{1}{2}L\right)$$

$$V_x = \left(\frac{wL}{2}\right) - wx$$

$$V_{max} = \frac{wL}{2} \quad (x = 0)$$

$$y = \left(\frac{5WL^3}{384EI}\right)(max)$$

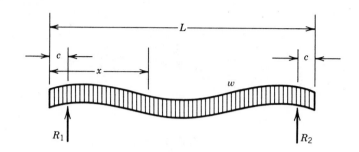

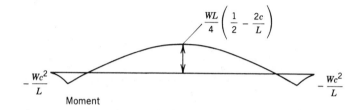

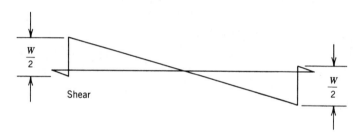

$$R_1 = \frac{W}{2} = \frac{wL}{2}, \qquad R_2 = \frac{W}{2} = \frac{wL}{2}$$

$$M_x = \left(\frac{Wx}{2}\right)\left[1 - \left(\frac{c}{x}\right) - \left(\frac{x}{L}\right)\right] \qquad (x > c)$$

$$M_x = -\left(\frac{Wx^2}{2L}\right) \qquad (x \leqq c)$$

$$M_{max} = \left(\frac{WL}{4}\right)\left[\frac{1}{2} - \left(\frac{2c}{L}\right)\right], \qquad c \leqq \left[\frac{(\sqrt{2}-1)}{2}\right]L$$

$$V_x = \left(\frac{W}{2}\right) - wx \qquad (x > c)$$

$$V_x = -wx \qquad (x \leqq c)$$

Source: Theodore Baumeister and Lionel S. Marks, eds., *Standard Handbook for Mechanical Engineer.* (New York: McGraw-Hill Book Company, 1967), pp. 5-29–5-33.

REFERENCES

[1] Kuske, A., and G. Robertson, *Photoelastic Stress Analysis*, Wiley, New York, 1974.

[2] Blodgett, O. W., *Design of Weldments*, The James F. Lincoln Arc Welding Foundation, Cleveland, Ohio, 1972.

[3] Den Hartog, J. P., *Strength of Materials*, McGraw-Hill, New York, 1949.

[4] Juvinall, R. C., Engineering Considerations of *Stress, Strain, and Strength*, McGraw-Hill, New York, 1967.

[5] Photoelastic Stress Analysis, Bulletin SFC-200, Photoelastic, Inc., Malvern, Pa.

[6] *Zytel Nylon Resins Design Handbook*, E. I. Du Pont de Nemours & Company, Wilmington, Del., 1972.

[7] Popov, E. P., *Mechanics of Materials*, Prentice-Hall, Englewood Cliffs, N.J., 1952.

[8] Den Hartog, J. P., *Advanced Strength of Materials*, McGraw-Hill, New York, 1952.

[9] Kaye, S. W., "Building Strength into Brackets," *Mechanical Details for Product Design*, D. C. Greenwood, ed., McGraw-Hill, New York, 1964, pp. 108–109.

[10] Caswell, J. S., "Design of Parts for Conditions of Variable Stress," *Product Engineering Design Manual*, D. C. Greenwood, ed., McGraw-Hill, New York, 1959, pp. 302–303.

[11] Timoshenko, S. P., and J. N. Goodier, *Theory of Elasticity*, McGraw-Hill, New York, 1970.

[12] Hall, A. S., A. R. Holowenko, and H. G. Laughlin, *Theory and Problems of Machine Design*, Schaum's Outline Series, McGraw-Hill, New York, 1961.

[13] Heath, W. W., "Ten Different Types of Splined Connections," *Product Engineering Design Manual*, D. C. Greenwood, ed., McGraw-Hill, New York, 1959, pp. 100–101.

[14] "Low Cost Methods of Coupling Small Diameter Shafts," *Product Engineering Design Manual*, D. C. Greenwood, ed., McGraw-Hill, New York, 1959, pp. 96–97.

[15] *Handbook of Adhesive Bonding*, C. V. Cagle, ed., McGraw-Hill, New York, 1973.

[16] Lindberg, R. A., and N. R. Braton, *Welding and Other Joining Processes*, Allyn & Bacon, Boston, 1976.

[17] Erickson, W., "Cutting Drive Size and Cost with Narrow V-Belts," *Machine Design*, Vol. *51*, January 25, 1979, pp. 116–119.

[18] "Typical Welded Bases and Pedestals-I," *Mechanical Details for Product Design*, D. C. Greenwood, ed., McGraw-Hill, New York, 1964, pp. 328–329.

[19] Fisher, O. W., "Built-up Welded Constructions," *Product Engineering Design Manual*, D. C. Greenwood, ed., McGraw-Hill, New York, 1959, pp. 334–337.

[20] "Rex Power Transmission and Conveyor Components," Catalog R76, Rexnord Inc., Milwaukee, Wis. 1976.

[21] "Sprockets and Roller Chains," Catalog 67, Cullman Wheel Co., Chicago, Ill., 1967.

[22] Hemp, W. S., *Optimum Structures*, Clarendon Press, Oxford University Press, London, 1973.

[23] Roark, R. J., and W. C. Young, *Formulas for Stress and Strain*, McGraw-Hill, New York, 1975, pp. 516–517.

[24] Burr, A. H., *Mechanical Analysis and Design*, Elsevier, New York, 1981, p. 493.

[25] Timoshenko, S., and J. N. Goodier, *Theory of Elasticity*, McGraw-Hill, New York, 1951, pp. 372–382.

[26] Shigley, J., *Applied Mechanics of Materials*, McGraw-Hill, New York, 1976.

[27] Faupel, J., *Engineering Design*, Wiley, New York, 1964.

[28] Scipio, A., *Structural Design Concepts*, NASA SP-5039 National Aeronautics and Space Administration, Washington, D.C., 1967.

[29] Paradise, R., and G. Church, *Problems in Strength of Materials*, Blackie and Son, London, 1959.

[30] Roark, R., *Formulas for Stress and Strain*, McGraw-Hill, New York, 1965.

[31] Shanley, F., *Strength of Materials*, McGraw-Hill, New York, 1957.

[32] Johnson, K. L., "Non-Hertzian Contact of Elastic Spheres," *The Mechanics of Contact Between Deformable Bodies*, A. D. de Pater and J. J. Kalker, eds., Delft University Press, Netherlands, 1975, pp. 26–40.

[33] Rothbart, H., *Mechanical Design and Systems Handbook*, McGraw-Hill, New York, 1964.

[34] Muvdi, B., and J. McNabb, *Engineering Mechanics of Materials*, Macmillan, 1980.

[35] Zurn, F. W., "Crowned Tooth Gear Type Couplings," *Iron and Steel Engineer*, August 1957.

[36] Huebner, K., *The Finite Element Method for Engineers*, Wiley, New York, 1975.

[37] Zienkiewicz, O., and Y. Cheung, *The Finite Element Method in Structural and Continuum Mechanics*, McGraw-Hill, London, 1967.

[38] Timoshenko, S., *Strength of Materials Part II*, D. Van Nostrand, New York, 1941.

[39] Marshek, K. M., and M. A. Rosenberg, "Designing Lightweight Frames," *Machine Design*, Vol. 47, May 15, 1975, p. 88.

[40] Marshek, K. M., and H. H. Chen, "Numerical Solution for Contact Pressure Distributions in Misaligned Self-Lubricated Journal Bearings," SAE Paper No. 841125, September 1984.

[41] Radzimovsky, E. I., "Stress Distribution and Strength Condition of Two Rolling Cylinders Pressed Together," University of Illinois at Urbana-Champaign Engineering Expt. Sta., Bull. 408, Vol. 50, No. 44, 1953.

[42] Conway, H., and K. Farnham, "Contact Stresses Between Cylindrical Shafts and Sleeves," *International Journal of Engineering Science*, Vol. 5, pp. 541–554, Pergamon Press, England, 1967.

[43] Conway, H., S. Vogel, K. Farnham, and S. So, "Normal and Shearing Contact Stresses in Indented Strips and Slabs," *International Journal of Engineering Science*, Vol. 4, pp. 343–359, Pergamon Press, England, 1966.

[44] Chow, F., P. Engel, D. Heath, S. Lawphongpanich "Contact Stress and Wear for Type Characters," *IBM Journal of Research and Development*, Vol. 22, No. 6, pp. 658–667, International Business Machines, November 1978.

[45] Marshek, K. M., "On the Analyses of Sprocket Load Distribution," *Mechanism and Machine Theory*, Vol. 14, No. 2, pp. 135–139, Pergamon Press, New York, 1979.

[46] Gerbert, G., H. Jonsson, V. Perrson, and G. Stensson, "Load Distribution in Timing Belts," *Journal of Mechanical Design*, Transactions of the ASME, Vol. 100, No. 2, pp. 208–215, 1978.

[47] Miller, D. L., K. M. Marshek, and M. R. Naji, "Determination of Load Distribution in a Threaded Connection," *Mechanism and Machine Theory*, Vol. 18, No. 6, pp. 421–430, Pergamon Press, England, 1983.

[48] Shames, I. H., *Engineering Mechanics Vol. I, Statics*, Prentice-Hall, Englewood Cliffs, N.J., 1966.

[49] Shigley, J. E., and L. D. Mitchell, *Mechanical Engineering Design*, 4th Ed., McGraw-Hill, New York, 1983.

[50] "Fort Worth Sprockets and Roller Chain," Catalog ES106, Fort Worth Steel and Machinery Company, Houston, Tex., 1967.

[51] Peterson, R. E., *Stress Concentration Factors*, Wiley, New York, 1974.

[52] Conry, T. F., and A. Seireg, "A Mathematical Programming Method for Design of Elastic Bodies in Contact", *Journal of Applied Mechanics*, Trans. of the ASME, Vol. 38, pp. 387–392, 1971.

INDEX